Transport and Energy Processes 2015

Core Programming Area at the 2015 AIChE Annual Meeting

Salt Lake City, Utah, USA
8-13 November 2015

ISBN: 978-1-5108-1875-0

Printed from e-media with permission by:

Curran Associates, Inc.
57 Morehouse Lane
Red Hook, NY 12571

Printed by Curran Associates, Inc. (2016)

For permission requests, please contact AIChE
at the address below.

AIChE
120 Wall Street, FL 23
New York, NY 10005-4020

Phone: (800) 242-4363
Fax: (203) 775-5177

www.aiche.org

Additional copies of this publication are available from:

Curran Associates, Inc.
57 Morehouse Lane
Red Hook, NY 12571 USA
Phone: 845-758-0400
Fax: 845-758-2634
Email: curran@proceedings.com
Web: www.proceedings.com

TABLE OF CONTENTS

432880 Analysis of a Nearly Isoentropic High Efficiency Pumped Heat Electricity Storage Process

Abstract:

Pumped heat electricity storage (PHES) is a recently proposed competitive energy storage solution for large scale electrical energy storage (EES). It is especially valuable for regions where specific geological structures are not available. The performance of PHES depends on two factors: the operations of turbomachines and the thermal storage system. The former is characterized by pressure ratio, polytropic efficiency and gas heat capacity ratio. The latter contains the parameters of heat regenerators that can be summarized into two dimensionless numbers: length Λ and step time π. The overall process operation can be described by temperature difference representing the energy stored per unit heat capacity, the storage bed utilization ratio and the turn-around efficiency. Exponential matrix solutions are obtained for a discretized heat transfer model of a typical pumped heat electricity storage process. Using the cyclic steady state and transient state solutions, we are able to analyze how dimensionless length and step time affect the storage bed utilization ratio as well as the turn-around efficiency. This model provides basic guidance for further development of such processes.

426247 CFD Simulation of CO2 Regeneration of Carbonated MgO Based Solid Sorbent in a Carbon Capture Process

Abstract:

There is an ever-increasing global pressure to reduce the emissions of CO_2, by capturing carbon from large stationary sources such as fossil fuel power plants. Currently, commercially available carbon capture technologies requires a relatively large volume of low-pressure steam from the power plant's steam cycle for regeneration that decreases the net power produced by the plant. Novel solid sorbents are believed to be an energy efficient solution to the problem. Thus in this work a novel regenerative process for high temperature CO_2 removal from Syngas using a MgO-based sorbent has been proposed. One of the major advantages of this high temperature sorbent is the elimination Syngas cooling/reheating that is necessary for current CO_2 separation systems. We have also shown that the circulating fluidized bed (CFB) concept ensures continuous CO_2 removal processes based on the dry sorbent concepts. However, one of the challenges in the way of deployment of this promising technology, among other novel technologies, is the fact that the majority of them are still in the lab or bench scales and, to be successfully scaled-up, a powerful tool such as Computational fluid dynamics (CFD), is needed to bridge the gap between lab/bench scale and large scales needed for demonstration. On the other hand, despite of the long history of CFB operations, complete understanding of the origin and nature of the inherently complex flow patterns observed in these devices is still lacking and beside experimental studies, advanced and rigorous numerical modeling is essential to shed light on the complex behavior and flow structure in these systems.

In this study first a high capacity, highly reactive MgO - based regenerative sorbent was developed for the purpose of CO2 capture. Then a series of sorption and regeneration experimental runs were performed and a second order shrinking core reaction model for regeneration was developed based on our experimental data and was incorporated in our CFD simulation of CO_2 regeneration process. Our and literature experimental data on fresh, carbonated, and regenerated sorbent particles have shown that the sorbent particle size distribution and density do not change significantly during the carbonation and regeneration processes. Therefore constant particle size and density were assumed for all of our CFD simulations. In addition simulation of the hydrodynamics of the entire circulating fluidized bed system loop were performed to obtain realistic values for the solid phase circulation rate to be used in simulation of the regeneration process. Then two dimensional both batch continuous and three dimensional continuous CO_2 regeneration process operating at the bubbling flow regime along with second order shrinking core reaction model was simulated using our CFD model. Our calculated CO_2 regeneration and conversion are in line with expected experimental expectation.

Pneumatic Conveying in High and Low Temperatures

George E. Klinzing

AIChE Annual Meeting Salt Lake City, Utah

For the John Chen Memorial Session

Novemeber 8-14, 2015

Abstract

Pneumatic conveying generally takes place at ambient conditions which can vary depending on the locale. Have the transport take place at extremes in temperatures is not that common but has been applied successfully in a number of industrial operations. For high temperatures the conveying of sponge iron sphere at 700 F and the conveying of ice to cool deep mines are the two extremes that are noted. Using steam as the conveying gas is another application of pneumatic conveying. The properties of the material conveyed often changes character as the temperature changes opening up a whole series of challenges. The application of another operation such as drying of the conveyed material during transport has been seen in the food industry but also can be applied to other industries. Looking at the concept of process intensification shows promise for these dual operation possibilities.

Sponge Iron Spheres (1.2 cm diameter) Conveying for the Direct Reduction of Iron Ore Process

The design on the system for conveying sponge iron spheres was developed at HYLSA, a steel production facility in Monterrey, Mexico. Since successful execution of this design, two additional installations has been successfully designed in other parts of the world. Flores and Solt () as the leader of the design team described the process for this high temperature application. The sponge iron spheres are very abrasive while also having great chemical reactivity with air with a range of temperatures from 35 to 900 C. Heat loss for the system was to be minimized and of course cost was an important factor. The conveying gas was a mixture of hydrogen and carbon monoxide. The material was directly fed to a reactor without a pressure lock.

In the operation and testing it was found that transition zone between dense and dilute was flat and fuzzy. The saltation point was not a sharp transition since the particles did roll in the pipe. The sponge iorn forms a laminated crust that can be abrasive to the pipe. Above 620 C the particles show a plastic flow region deforming to a smaller layer of iron on the pipe surface. The alloy pipelines chosen for the design has surpassed 16 years of operation.

Scale-up of the design was done from a 4 inch pipe to 12 inch pipe with a distance of 100 m to 300 m.

Selection of high temperature valves, flanges and bolts had to be considered. The whole design process including the pilot plant took 6 years to develop.

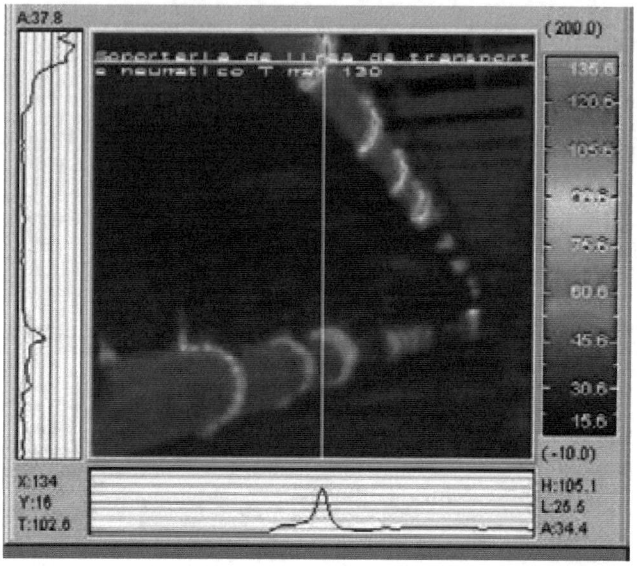

Figure 1 – HLYSA temperature profile for high temperature conveying line

Pneumatic Transport of Coal by Steam

Steam can be used as the transport gas in a power plant where high-pressure steam is available (Wildman et al). The steam can provide the transport gas and also a source of oxygen. Steam is an abundant resource in a power plant. Another possible application of steam as the conveying gas is in coal gasification. A loading ratio up to 10 was used to probe the feasibility of the conveying in a 0.01 m diameter pipe. Using the standard pressure loss calculations with the operation at isothermal conditions Figure shows the results. Employing a polytropic condition for the temperature showed good agreement between the data and the analysis.

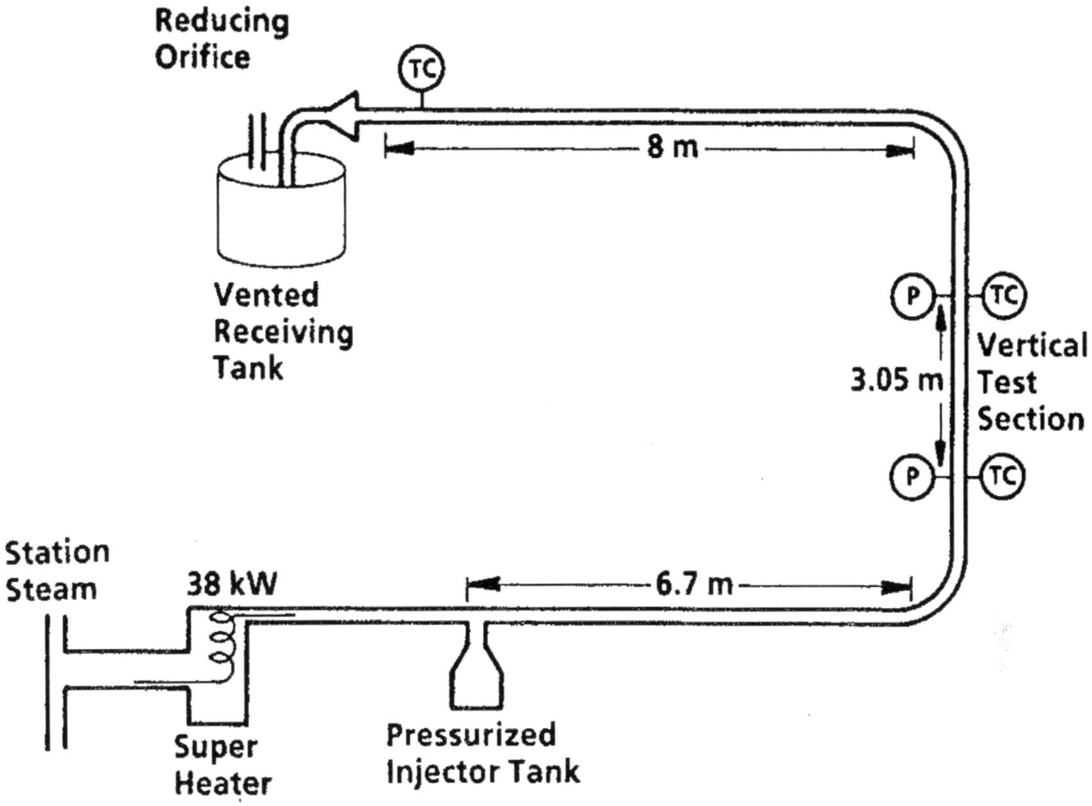

Figure 2 – Steam Conveying Rig

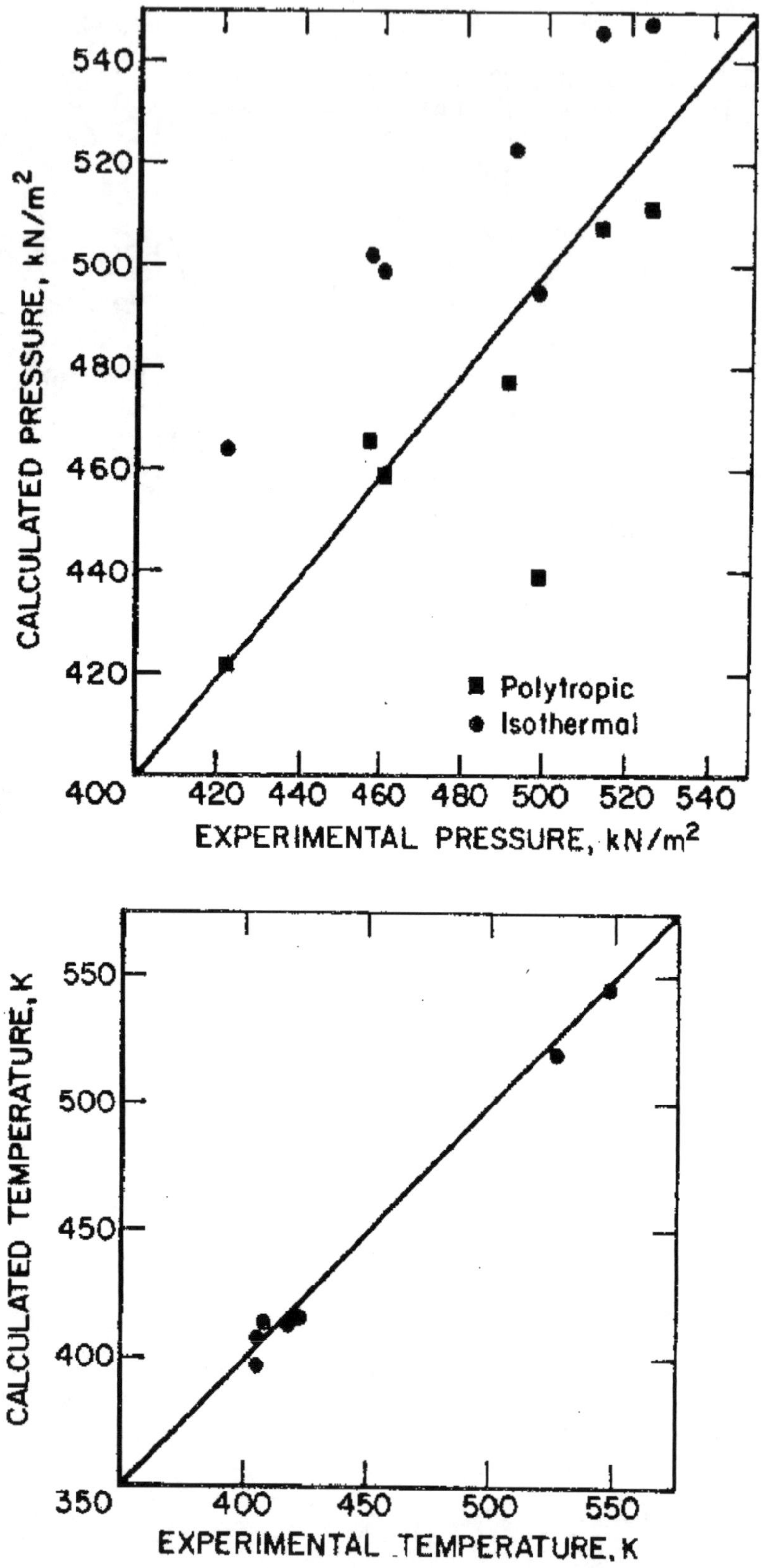

Figure 3 Pressure drop and temperature of steam conveying system

Development of a High Temperature Flowmeter

A number of flow measurements of solids have been done in our laboratory using the principle of cross-correlation of two signals roughly 1 to 3 feet apart in horizontal and vertical pneumatic conveying. The principle is to use the electrostatic signal generated in a non-conductive section of pipe preceded and followed by a conductive section of pipe use as an electrode to collect the electrostatic signal generated. In experimental testing of the conveying and resistance of ceramic line pipe at high temperatures a similar type of experimental arrangement was implemented. The ceramic liner serviced as the non-conductive section of the piping with a small ring of conductive metal separating the con-conductive section. The operation took place at 1000 F, (Klinzing et al.)

The high temperature particle velocity meters is comprised of a pair of spaced apart conducting portions having disposed on each of their upstream edges a first and second non-conducting portion. The bombardment of the partiles on the conducint and non-conductin potions creates charged partles in the pneumatically transported material. These particles dischange an electric current through the conducint poitions to respective conductors and then to respective signal generating means. The singal generating means convert the electric dischamge to an electronic signal. The electronic signalsfrl each of the signal generating means are fed into a calculating means which calculate the particle velocity based on the respective electronic signals. In another application can be applied to a ceramic or metal pipeline. This device has been shown to be operative at temperatures over 1000° F.

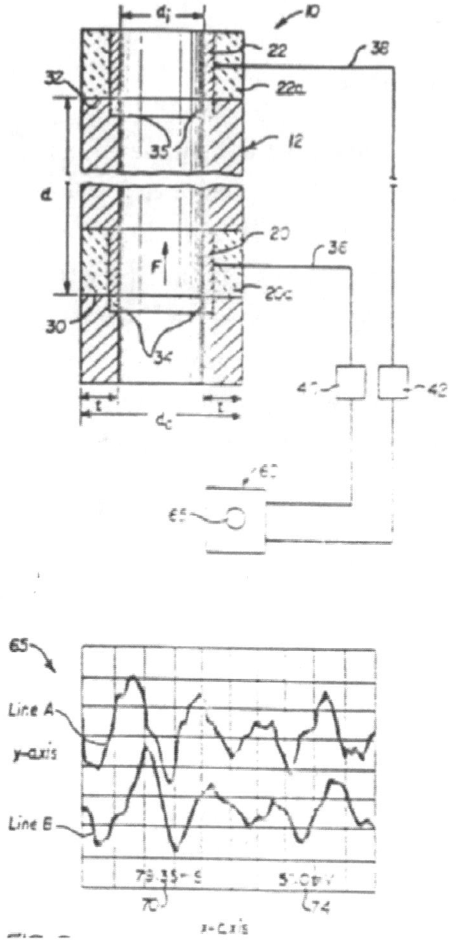

Figure 4 – High temperature measurement device for particle velocities

General Temperature Consideration in a Pneumatic Conveying System

When operating and designing a pneumatic conveying system on must recognize that colder air entering a system from external sources has a higher density and thus smaller volume than warm air that may be used for conveying. If the blower is not given sufficient time to warm up, the outset temperature of the gas could be low leading to lower than expected volumetric flow rates. One should run the blower for at least 15 minutes to warm up before beginning the conveying operation.

If any portion of uninsulated or non-heat-traced conveying line is exposed to ambient conditions where the night time temperatures falls by at least 20 F or below the dew point one will be faced with condensation which can affect the product being conveyed and could lead to an upset or even blockage of the pipe. It is advisable to flow dry air or ambient air through the system long enough to evaporate the moisture can may have condensed in the line. One can also keep the line insulated and traced whenever a pipe goes outdoor. Continuously purging the system to prevent the ingress of humid air may also be an option.

Material Affected by High Temperature Changes in Pneumatic Conveying

Analysis of wear of a ceramic pipe by particles of slag from a wet bottom boiler produced some very unexpected results when the temperature were at 500 to 1000 F. This slag material is very abrasive having a sharp angular shape with a glassy appearance. The materials was chosen because of its similarity to many coal related materials. A high temperature vertical test facility was designed having a ceramic lining. As the conveying was implemented one found an increase in the degree of wear of the piping liner. As the temperature reached 450 F one finds that the wear rate decreases dramatically. Figure () shows this behavior. On further analysis of the slag it was found now to have a more rounded shape and be angular any more. At high temperatures these particles softened and thus further reduced the erosion rate. Impact of these particle with the wall are seen in the cartoon presented in Figure (), (Borzone et al.).

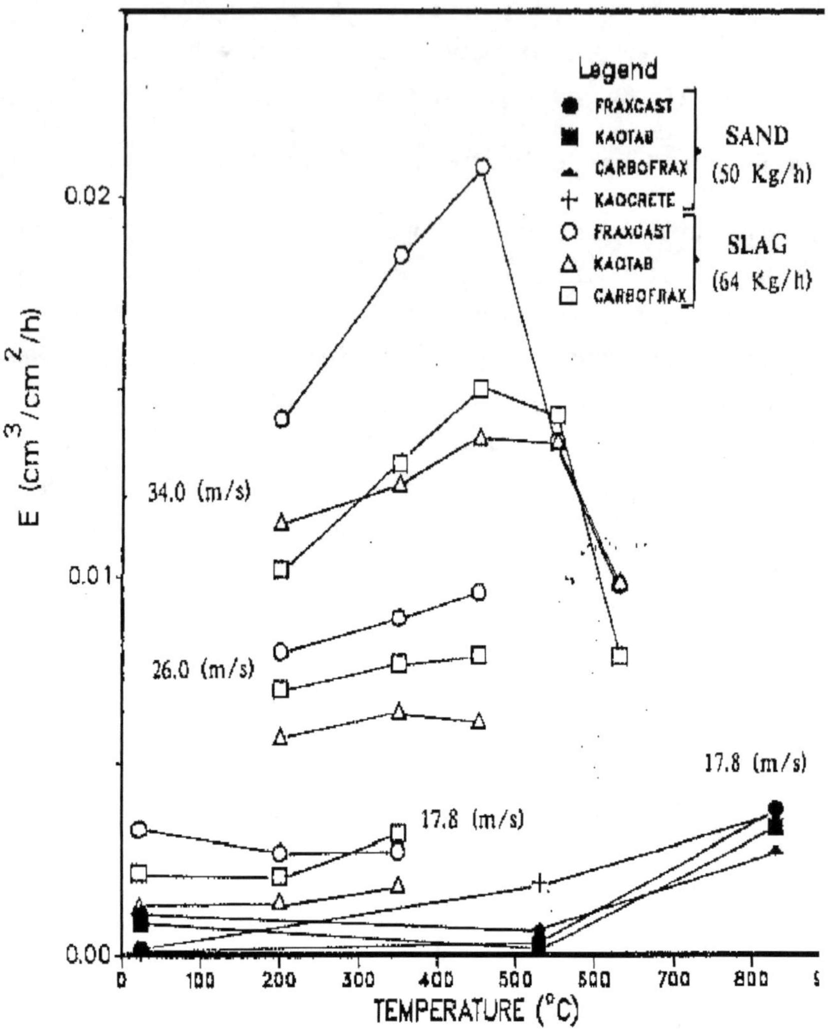
Legend
FRAXCAST
KAOTAB
CARBOFRAX
KAOCRETE
FRAXCAST
KAOTAB
CARBOFRAX
SAND
(50 Kg/h)
SLAG
(64 Kg/h)
34.0 (m/s)
26.0 (m/s)
17.8 (m/s)
17.8 (m/s)
0.02
0.01
0.00
E (cm³/cm²/h)
0
100
200
300
400
500
600
700
800
TEMPERATURE (°C)

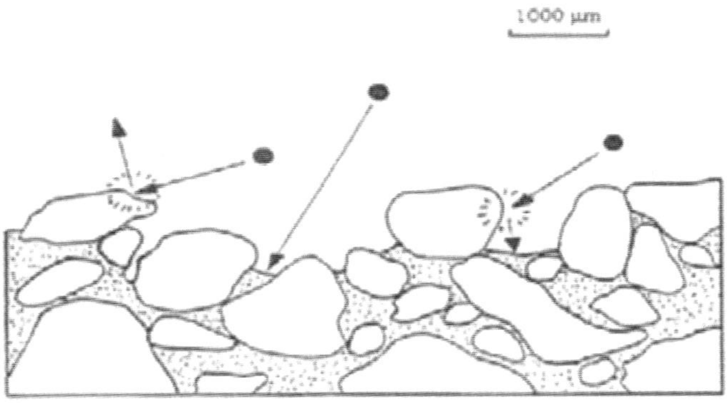

Figure 5 – High temperature conveying of slag wear behavior

Coal Particles Behavior in a Bend

The behavior of bends in feeding coal to a blast furnace can be problematic in that these bends tend to block the flow causing reduced coal flow to the blast furnace. To explore this phenomena Venkatasubramanian et al. experimentally implement studies on a short radius bend, long radius bends and a T-bend to elucidated their behaviors. The coal used in the study and in the blast furnace is a high volatile coal that can release volatile materials when the temperature of the coal is increased. The flow conditions and solids loading closely mimicked that of a blast furnace. For test runs of 90 minutes at 20 minutes deposits were found in 20 minutes in the long-radius bend became 50% blocked with coal at the end of the run.

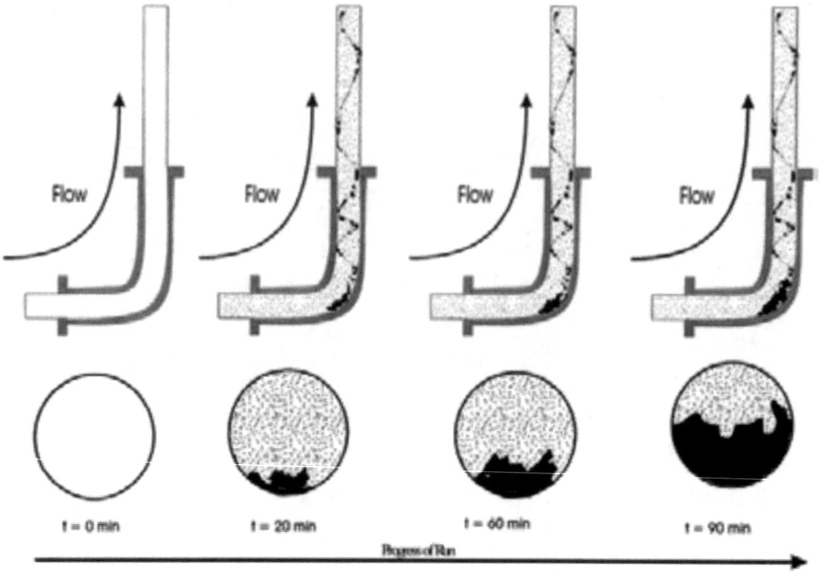

Figure 6 – Long distance bend deposits of coal

With the short radius bend the streaky deposits were seen first to form becoming wide downstream of the curse of the bend. There were also deposits at the secondary impact point of the coal. Both the short radius bend and the T-Bend showed roping in the vertical distance beyond on the bend which was not observed in the long-radius bend.

The high volatile coal reacts to the impact of the particles with the wall and the rise in temperature of the point where the particles devolatilize and become sticky attracting more particles and building a deposit. This is more severe in long radius bends which have a longer contact time with the wall than the short-radius and T-bends.

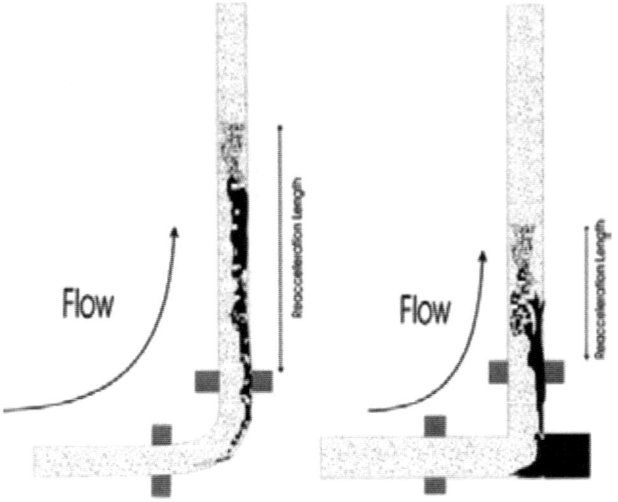

Figure 7 – Short-radius bend and T-bend coal deposits

Collection of Sodium from Combustion of Coal with a Getter Material

The combustion of coal can release a number of different components in the off gases. One of concern in the environmental and equipment operation is the release of sodium from the coal. In a study carried out we employed a getter material, alumina BW 2500, injected into the coal combustion region of a flat flame combustor flowing in a downward vertical direction. Figure () shows the experimental setup. Alumina as well as other possible getter materials for the sodium were employed in the study. It was found from the photomicrographs that the alumina could be employed as getter for sodium, (Shannon, Martello).

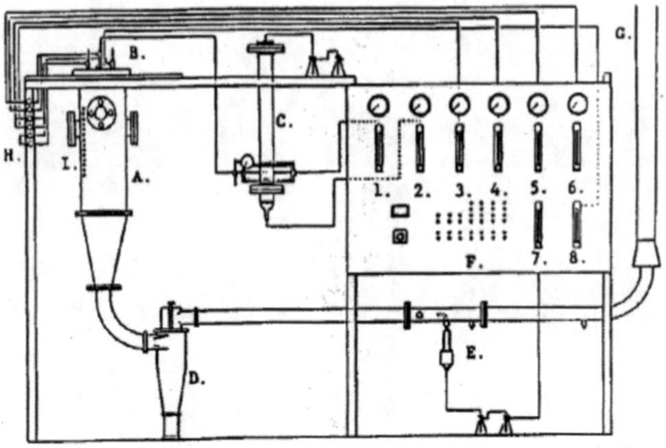

Burner Components
A. 5" Burner Chimney
B. Burner Feed Assembly
C. Fluidized Bed Coal Feeder
D. Cyclone
E. Anderson Impactor
F. Control Panel
G. Exhaust
H. Fine Metering Valves
I. Thermocouple Ports

System Feed Streams
1. Coal Carrier Air
2. Fluidizing Air
3. Annular Burner Gas
4. Annular Burner Air
5. Inner Burner Gas
6. Inner Burner Air
7. Vacuum
8. Cooling Water

Figure 8 – Rig for testing getter capacity for sodium in coal combustion

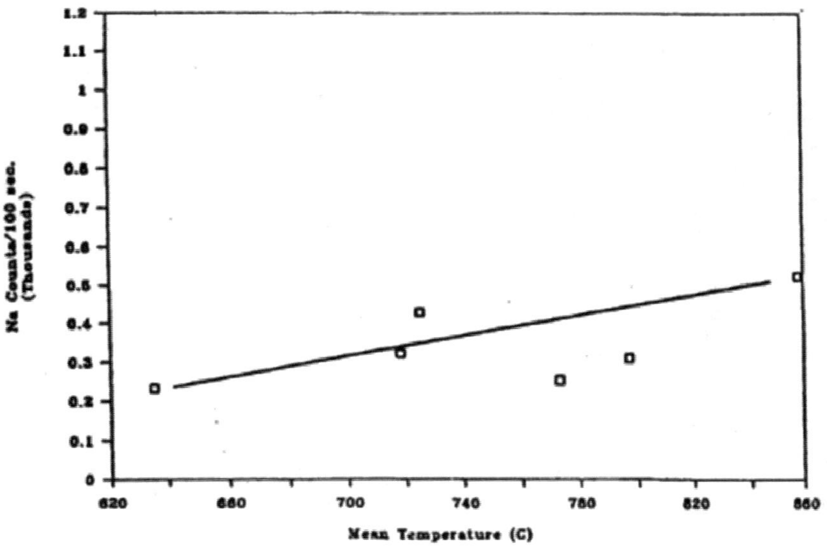

Figure 9 – Sodium capture plot as a function of temperature of combustion

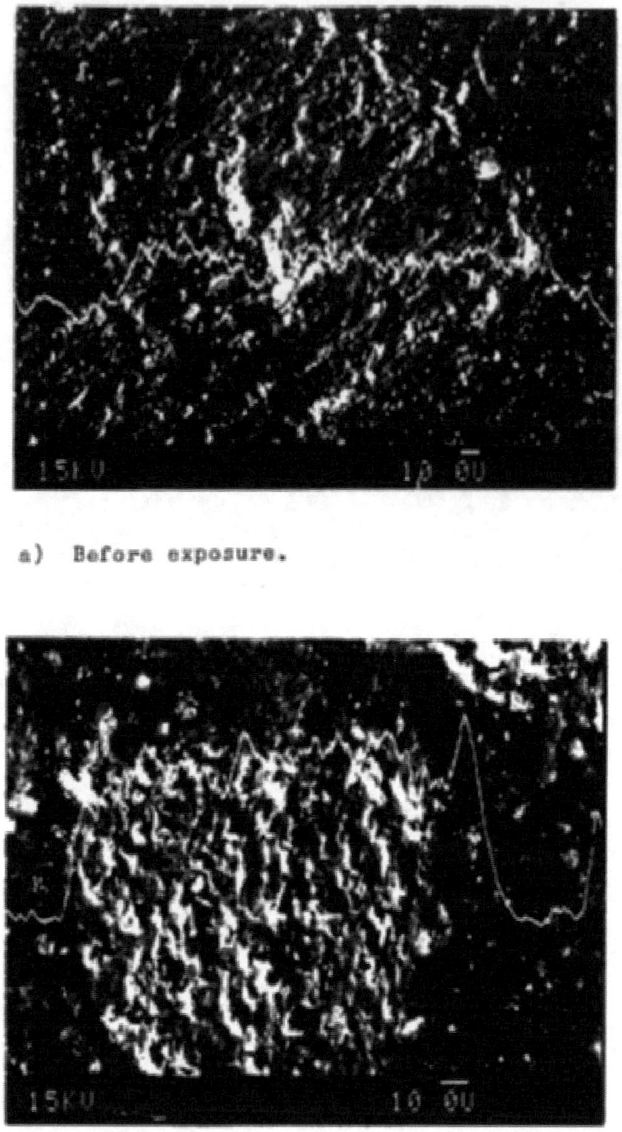

a) Before exposure.

b) After exposure.

Figure 10 – BG2500 alumina WDS Line scan for sodium

Pneumatic Conveying of Ice to Deep Mines

Shear has shown that it is possible to conveying ice to a deep gold mine for cooling and is more economical that cooling water alone. Ice Conveying and use of the latent heat of fusion is sufficiently large to allow a reduction in water pumping rate to 1/5 of that in a conventional system for the same heat removal rate.

At a rate of 43 kg/s of ice or 3700 T/day giving up 18 MW of cooling can be achieved. Ice delivered to 2407 m underground with another 1500 m for underground horizontal conveying. In a 136 mm diameter pipe with max icechip size of 36 mm was achieved. The piping for the conveying was plastic permitting less adhesion of the ice with the metal surface and reducing the friction of the moving ice. The solids friction factor was correlated with the Froude number of the flow to yield

$$\Lambda_s = 0.084\ Fr^{-0.5}$$

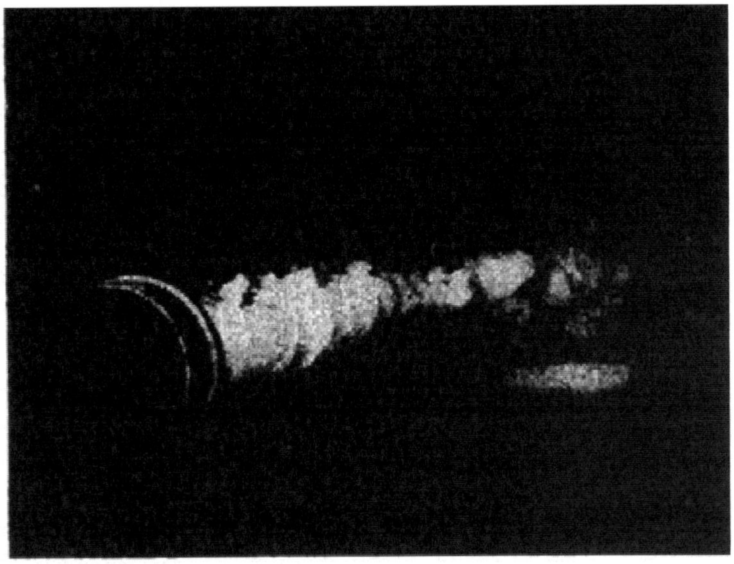

Figure 11 – Loose ice slug emerging from the pipeline

Conveying and Drying or other Process Possibilities

Having another unit operation take place as conveying is implemented is something that has happened quite a bit in the food processing industry with the drying of foodstuff as the material is being transported. Reviewing the literature one find a German patent in 1910 referring to such an operation. Having mass and momentum transfer simultaneously is not uncommon. The term process intensification comes to mind when multi operations are proposed as a means of optimizing a system as well as saving energy in the process. Such operations as granulation and conveying is another possible application of simultaneous operations.

References

Borzone, L.A., Klinzing, G.E., Yang, W.-C. "Erosion at high temperature in pneumatic transport," Proceedings of the Joint ASME-Chinese Academy Meeting on Multiphase Flow, August, 1987.

Flores, M. and Solt, P., "High Temperature Pneumatic Conveying," Powder and Bulk.com, Ask Joe! Column

Klinzing, G.E., Borzone, L.A., Yang, W-C.High Temperature Particle Velocity Meter and Associated Method, U.S. Patent No. 5,022,274 (1991)

Martello, D.V., "Adsorption of sodium chloride vapor on porous silica and alumina powders," M.S. Thesis, University of Pittsburgh (1987)

Shannon, S.L., "The use of solids collectors for the control of sodium in advanced coal conversion processes," M.S. Thesis, University of Pittsburgh (1986)

Shear, T.J., "Pneumatic conveying of ice particles through min-shaft pipelines," Powder Tech, 85 (1995) 203-219.

Venkatasubramanian, S., Tashior, H., Klinzing, G.E., Mykelbust, K., "Solids flow behavior in bends: assessing fine solids buildup, " Powder Tech, 113 (2000) 124-131.

Wildman, D.J., Ekmann, J.M., Mathur, M.P., Klinzing, G.E., "Pneumatic transport of coal by steam, " Chemical Engineering Symposium Series 225, Vol. 83, (1987) 58-63.

432083 High Temperature Thermal Energy Storage Using Phase Change Materials

Abstract:

The objective of this work is to establish methods for storage of thermal energy at high temperatures applicable in concentrating solar power plants (CSPs). Temperatures up to 600°C are considered using encapsulated phase change materials (EPCMs), where heat transfer fluid (HTF) from the solar collector would pass through the storage system embedded with EPCM capsules.

$NaNO_3$ and a eutectic compound of NaCl–$MgCl_2$ are selected as phase change materials (PCM), which are tested in a special calorimeter at temperatures below and above their melting points to determine their sensible and latent heat storage capabilities as well as stability and compatibility with the encapsulation material over multiple cycles of melting and solidification.

Considering the rate of melting and solidification inside the EPCM, optimum diameter of cylindrical capsules are found to be between 50 to 100 mm. Capsules with 76 mm diameter and 260 mm length, containing phase change material (PCM) are fabricated and installed in a pilot-scale thermal energy storage (TES) system for performance tests, where compressed air is used as heat transfer fluid. Tests with ten rows of EPCM capsules successfully demonstrate the ability to transfer thermal energy to and from the transport fluid, achieving energy storage and retrieval in multiple charging and discharging cycles.

A mathematical model has been developed for the test section with EPCM capsules and its predictions are found to agree with experimental measurements within 7% discrepancy in stored energy. The dynamic performance of charging and discharging rates are also well predicted by the model, giving confidence for engineering design capabilities in future applications using EPCMs for thermal energy storage.

432117 Entrainment Calculations Via Bubble Dynamic Modeling for Applications at Elevated Pressure and Temperature

Abstract:

Calculations of entrainment rate from an operating gas-solid fluidized bed have been problematic even under ambient operating conditions. Entrainment correlations available in the literature were mostly developed through multiple-regression of selected dimensionless groups employing limited database in term of particle properties (size, density, and distribution), bed diameter, bed height, distributor plate characteristics, and operating conditions. Thus the entrainment rates calculated from those models can differ by several orders of magnitude. Their extrapolation to elevated pressure and temperature is doubtful. A new mechanistic entrainment model has been developed by considering the bubble dynamics of a gas-solid fluidized bed. The development of bubbles is followed from their formation at the distributor plates of different designs, their growth through bubble coalescence, and their eruption at the bed surface. The solids carried in the bubble wake are assumed to be ejected into the freeboard forming the primary particles to be entrained. The actual entrainment rate is, however, based on the freeboard design (height and expanded section). The entrainment rate calculated from the model was compared to a proprietary data point estimated from an operating commercial unit operating at elevated pressure and moderate temperature.

415439 The Travelling Fluidized Bed Project: Working Together

Abstract:

John Chen had a strong attachment to "challenge problems" to evaluate the ability of models and CFD codes to accurately predict complex fluidization phenomena. He was instrumental in introducing challenge exercises at several fluidization conferences. He was also committed to the Fluidization community working together. In this spirit, the Travelling Fluidized Bed project was conceived as a means of comparing different advanced experimental techniques in measuring key hydrodynamic properties by setting up a transportable gas-solid fluidization unit at the laboratories of normally competing organizations to take measurement that should yield similar or complementary results, under identical operating conditions. At the same time, the experimental results would build a unique database, in which several different properties (e.g. distributions of voidage, particle velocity, solids flux) are measured by different invasive and non-invasive techniques with group A and B particles, providing an indication of experimental error. This presentation will report on progress after operation of the unit in 7 locations in three countries and on some of the lessons learned from this unique exercise in having different members of the community work together.

419187 Pervaporative Enrichment of 1,3-Propanediol from Model Fermentation Broths By Hydrophobic Specialty Polymers

Abstract:

While 1,3 PD can be easily separated from the other by products of bacterial metabolism, the real challenge lies in enriching it from dilute concentrations in aqueous mixtures which amounts to almost 50 – 70% of the total production cost in fermentative 1,3-PD production. The hydrophilic character of 1,3 PD compounds the difficulty of purification. Conventional and reported separation techniques have had their limitations in due cost and energy consumption and hence applicability. Membrane separation processes offer a cheap and energy efficient method of upgrading the value of specific components typically produced in dilute mixtures. Although pervaporation is an energetically advantageous process. 1,3-PD cannot be enriched by materials conventionally used for pervaporation. The development of novel materials with a good cost performance balance is essential for the establishment of industrial separation procedures for 1,3-PD enrichment from dilute broths. The present work describes a study on the pervaporation enrichment of 1,3-propanediol from model fermentation broths using three hydrophobic polymer families – functionalized siloxanes, cyclohexylamine base methacrylates and imidazolium ionic liquid based methacrylates with progressively increasing separation efficiency. While the highest 1,3-PD permeabilities were seen with the siloxane polymers, the imidazolium IL based methacrylates exhibited the best separation factors both with respect to water and glycerol. The effect of glycerol on transport coupling was evaluated. The permeability and separation factors from the pervaporation experiments with these polymers were siloxanes were collated to arrive at a Robeson's upper bound of pervaporation of 1,3-propanediol from aqueous solution by polymeric membranes. The development of novel materials with a good cost/performance balance is central to the establishment of industrial separation procedures for 1,3-PD enrichment from fermentation broths. The removal of this process bottleneck may increase the feasibility of using waste industrial glycerol as a sustainable resource for producing 1,3-PD as a biorenewable alternative to a petroleum platform.

IMPLEMENTATION OF A CUSTOMIZED GAS-SEPARATION MEMBRANE MODEL INTO COMMERCIAL FLOWSHEETING SOFTWARE TO SIMULATE A HYBRID CO_2 REMOVAL PROCESS FOR OXIDATIVE COUPLING OF METHANE

Alberto Penteado[1,2], Erik Esche[1], Günter Wozny[1]
(1) Process Dynamics and Operations Group, Technische Universität Berlin, Berlin, Germany
(2) CAPES Foundation, Ministry of Education of Brazil, Brasilia, DF, Brazil

Introduction

Ethene is one of the most produced basic chemical world-wide with a global production of 141 million tons in 2011 [1]. The oxidative coupling of methane (OCM) is a promising path for the conversion of methane into ethene, offering the possibility for utilizing natural gas, shale gas or biogas instead of oil as feedstock. It is a more direct route from methane to valuable chemicals compared to the Fischer-Tropsch process, which has synthesis gas as an intermediate [2]. The selective oxidation of methane to ethene can be seen below:

$$2CH_4 + O_2 \rightarrow C_2H_4 + 2H_2O$$

Due to the high temperatures required for catalyst activation in the OCM reaction (750 to 950 °C), products of non-selective oxidation such as carbon monoxide (CO) and carbon dioxide (CO_2) are also formed and should be removed in the separation section. The C_2 yield in the OCM reactor is not expected to exceed 28% [3], which highlights the need for an efficient downstream separation for an OCM process to become economically feasible.

The OCM process structure typically includes a reaction section, consisting of a reactor or reactor network; a compression section based on a multi-stage compressor with intermediate cooling and knock-out drums for water removal; a carbon dioxide removal section, commonly via an absorption/desorption system; and a distillation section, consisting of a demethanizer column and a C_2 splitter. This structure is reproduced in the block diagram in Figure 1 below.

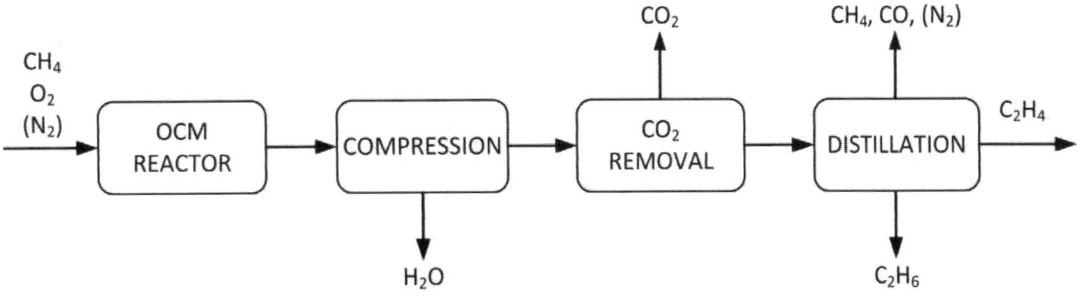

Figure 1 - Typical OCM process structure

The carbon dioxide removal is usually an energy-intensive step, since the amine solution used for the absorption needs to be regenerated in a stripping column by adding heat through the reboiler. Alternatively, a hybrid process can be employed. This consists of an upstream gas separation membrane section and a downstream absorption section. This concept has been successfully applied in a mini-plant in the Technische Universtiät Berlin, showing significant reduction in the specific energy required per kilogram of captured CO_2 [4]. In this contribution, the feasibility of applying this process at an industrial scale is assessed through simulations and preliminary cost estimations.

Process Concept

Amine solutions are often utilized for the sequestration of CO_2 in power plants or for natural gas sweetening. A similar approach is used for the OCM process. The OCM reactor outlet passes through the compression and water removal section and enters the CO_2 removal section, where it is fed to a packed column in counter-current flow with a 30 wt% Monoethanolamine (MEA) acqueous solution. The carbon dioxide is dissolved and chemically bound, producing a CO_2-free gas stream at the top and a CO_2-rich amine solution at the bottom. The rich amine solution is than flashed, pumped, and pre-heated before entering the stripper column, within which it is regenerated. A gas stream consisting mainly of CO_2 and water vapor is produced at the top of the stripper column, whilst a lean amine solution is produced at the bottom. The latter is then pumped, cooled, and recycled back to the absorption column.

In the proposed hybrid process, a gas separation membrane unit or unit cascade is added upstream of the absorption section. The carbon dioxide is partially removed in the membrane section, reducing the amine flow rate and reboiler duty in the absorption section for a fixed degree of CO_2 capture. This concept is represented in Figure 2, wherein a two-stage stripping cascade is employed.

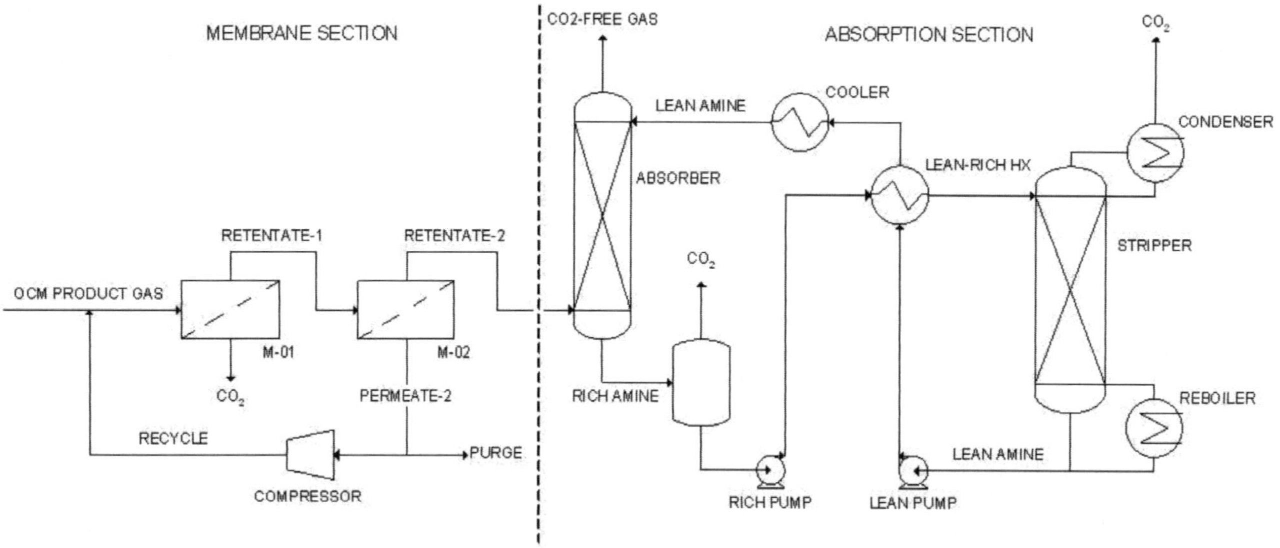

Figure 2 - Process flow diagram for the hybrid CO_2 capture process

Membrane units consist of a series of circular flat sheet membrane envelopes with segment cuts disposed in series inside a metal casting containing an inner tube for the permeate flow as seen in Figure 3. Both Polyimide (PI/Matrimid) and Poly-(ethylene oxide) (PEO) membranes are considered in this contribution. The first has higher selectivity to CO_2, whilst the second allows for a larger flow through the membrane. Membrane materials and modules are produced by coworkers at Helmholz-Zentrum Geesthacht Centre for Materials and Coastal Research, Germany, as described in [5].

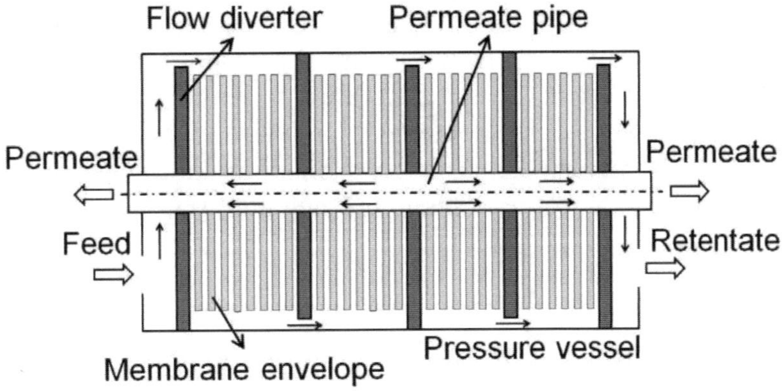

Figure 3 – Representation of the envelope type membrane unit

Modeling

The global model is divided into a membrane section and an absorption section. The model for the membrane section is implemented using the web-based modeling environment MOSAIC [6]. MOSAIC uses a document-based modeling approach to facilitate the documentation step and the exchange and reuse of models across platforms. It allows for automatic code generation for the model solution in different programming languages and environments. A code for the solution of the membrane model in Aspen Custom Modeler® (ACM) is generated and the ACM model is later exported to Aspen Plus® as a unit operation for the simulations. The absorption section is modeled in Aspen Plus® using available models from the library and from the example file [7].

Membrane Section

Equations for modeling gas permeation membranes are presented in [8]. A one-dimensional solution-diffusion model is applied and discretized using orthogonal collocation on finite elements with fourth order Lagrangian polynomials on Radau roots. Basic modeling equations are presented below: molar balances on permeate and retentate sides (Equation 1), permeance calculation for the PI membrane according to the Arrhenius law (Equation 2) and permeance calculation for the PEO membranes according to the free-volume theory (Equation 3). Further detail on modeling equations and implementation is given in [4].

$$\frac{dn_i^P}{dz} = -\frac{dn_i^R}{dz} = L_i \cdot \left(f_i^R - f_i^P\right) \cdot h \qquad (1)$$

$$L_i^{PI} = L_i^{0,PI} \cdot exp\left(-\frac{Ea_i^{PI}}{R \cdot T}\right) \qquad (2)$$

$$L_i^{PEO} = L_i^{0,PEO} \cdot exp\left[-\frac{Ea_i^{PEO}}{R \cdot T} + \sigma_i^2 \sum_{i=1}^{NC} \left(\frac{1}{\sigma_i^2} \cdot m_i^0 \cdot \frac{f_i^R + f_i^P}{2} \cdot exp\left(m_i^T \cdot T\right)\right)\right] \qquad (3)$$

Permeabilities are obtained from co-workers at Helmholz-Zentrum Geesthacht Centre for Materials and Coastal Research, Germany [5]. Fugacitites are calculated using the Peng-Robinson equation of state on the retentate side, while ideal gas is assumed on permeate side.

Experiments carried within the mini-plant at the Technische Universität Berlin showed nearly isothermal conditions on both retentate and permeate sides, as well as negligible pressure drop and concentration polarization, therefore these effects are not taken into consideration. The model is initially used for the simulation and optimization of the mini-plant [4]. The model is then extended for the simulation of industrial size plants by using a scale-up by numbering-up approach. The inlet flow rate is divided by the design flow of the membrane module to obtain the required number of parallel modules. A single module is simulated and the resulting flows are then multiplied to the number of parallel modules to obtain the total retentate and permeate outlet flows.

Absorption Section

An unsymmetric Electrolyte NRTL model is used for the liquid phase and PC-SAFT equation of state is used for the vapor phase. Gaseous components, H_2, N_2, CH_4, C_2H_4, C_2H_6, CO, and CO_2 are selected as Henry components. The example file [7] is used as a starting point for the modeling of the absorption section. Implemented parameters from various sources are kept and the Henry constant for ethene in Monoethanolamine (MEA) is additionally estimated based on data from [9] to avoid an underestimation of ethene losses in the absorber. Columns are simulated using rate-based calculations with built-in correlations by [10] for mass transfer, by Chilton and Colburn [11] for heat-transfer and by [12] for the packing hold-ups. Kinetic parameters for the CO_2 absorption are taken from [13] and [14]. Plug flow is assumed for the vapor, while liquid is assumed to be well mixed. Column packing is Sulzer Mellapak 350X for the absorption column and 350Y for the stripping column. Other equipment (flash, pumps and exchangers) are simulated using the traditional MESH approach.

Simulations and Optimizations

Simulations are performed using Aspen Plus® version 8.6. All models are solved using the available numeric methods in the software. A sparse solver is used for the membrane modules, Broyden method is applied for converging tears in recycle calculations and SQP is used for sequential optimization of the membrane section.

Different reactor design and feeding policies for OCM are currently under investigation as described in [15] and [16]. As they impact the downstream separation processes drastically, an integrated approach is required to synthesize an efficient OCM process. In this contribution, multiple feed scenarios are considered to reflect possible reactor outlets of a plant producing 100 kton/year of ethene. All gas flows are assumed to be compressed, cooled, and dry, as if they were coming from the compression section. The feed scenario details are shown in Table 1 below.

Table 1 - Inlet gas conditions for the simulations

Feed Scenario	Temperature (°C)	Pressure (bar)	Molar Composition (%)			
			CO_2	C_2H_4	N_2	CH_4
I	35	10	27.4	15.4	34.4	22.8
II	35	10	9.0	9.0	16.7	65.3
III	35	10	24.5	4.5	8.0	63.0
IV	35	10	38.0	5.0	10.0	47.0

Scenario I reflects the use of nitrogen diluted flow in the reactor. Scenario II considers no dilution, but the use of a high methane to oxygen ratio instead. Scenarios III and IV consider carbon dioxide dilution, which has a positive impact on the C_2 selectivity and helps to control the high exothermicity of the OCM reaction network. For scenarios II, III, and IV, in which no nitrogen is present, other components (H_2, CO, and C_2H_6) are simulated as N_2, given that no permeances are available for those.

All simulations aim at the removal of 97% of the CO_2 content in the inlet gas. A base case consisting of a standalone absorption process is simulated for each scenario for the latter comparison with hybrid processes. Due to the size and complexity of the system, no actual optimization could be performed in the absorption section. Therefore, optimal design is carried in the form of extensive simulations and sensitivity analyses. A specific cost per kilogram of captured CO_2 is then calculated.

The membrane section is simulated separately. An objective function consisting of a simplified operational cost is used for its optimization. It comprises the compression cost, cost attributed to ethene losses and the cost for removing the remaining CO_2 through absorption at a fixed price calculated in the base case. Flowsheets containing a single PI membrane and a stripping cascade of PI and PEO membranes are considered. The decision variables are the membrane materials, the membrane areas and the recycle/purge ratio in the case of membrane cascades. Bad convergence for the recycle calculations limits the steps of the optimization solver. Many starting values are tested leading to different local optima. The downstream absorption section is then simulated only for the cases presenting a simplified operational cost lower than the base case.

Utility consumption and cost rates are calculated inside Aspen Plus®, allowing for the operational cost to be quickly assessed during the flowsheeting stage. Utility costs are set to their default values. Apart from the utilities, costs are also attributed to the ethene losses during the process, based on current market prices published on [17]. These costs are shown in Table 2 below.

Table 2 – Base costs for economic calculations

Type	Specific Cost	Unit
Cooling Water	2.12×10^{-7}	$/kJ
Electricity	0.0775	$/kWh
Steam	2.2×10^{-6}	$/kJ
Ethene Loss	1.00	$/kg

Cooling water is used in the condenser of the stripping column, amine solution cooler, and interstage coolers of compressors. Electricity is used in the amine circulation pumps and gas compressor. Steam is used in the reboiler of the stripping column. Ethene is lost via the permeate stream, the purge stream, and the absorber's bottom stream. The total operational cost (OPEX) is calculated as the sum of all applied utilities and the ethene losses.

Results

Table 3 shows the results of the cost estimations for the standalone absorption and for the hybrid processes for each feed scenario. It consists of the operational cost and the specific operational cost per ton of removed CO_2 and the membrane areas per module.

Table 3 - Operational costs for different feed scenarios

Scenario		OPEX (US$/h)	Specific OPEX (US$/ton$CO_2$)	PI area per module (m^2)	PEO area per module (m^2)
I	Standalone absorption	483.56	14.78		
	Single PI module	517.24	15.81	0.1	
	PI + PEO cascade	695.65	21.26	0.5	0.1
II	Standalone absorption	241.35	13.17		
	Single PI module	301.29	16.44	0.5	
	PI + PEO cascade	454.03	24.78	0.5	0.1
III	Standalone absorption	978.37	9.81		
	Single PI module	**924.75**	**9.27**	**1.0**	
	PI + PEO cascade	1132.32	11.35	2.3	0.1
IV	Standalone absorption	1456.66	10.46		
	Single PI module	**1353.27**	**9.72**	**2.5**	
	PI + PEO cascade	1548.33	11.12	2.0	0.1

For feed scenarios I and II, the cost for operating the hybrid process with a single PI module is higher than that of the standalone absorption. Even though the specific energy per kilogram of CO_2 removed and total utility costs are reduced, the total operational cost of the hybrid process is still higher because ethene losses are higher in the membrane process compared to the absorption. When utilizing CO_2 diluted streams, as in scenarios III and IV, the hybrid process with a single PI module becomes economically attractive. The higher CO_2 content in the feed stream increases the driving force for the separation in the membranes. Furthermore, the greater amount of CO_2 to be captured requires a larger amine recirculation rate, which also increases ethene loss in the absorption section up to the point that they become comparable to those in the membrane section. Economically feasible scenarios were found by applying a single PI module to remove carbon dioxide from feed scenarios III and IV.

The utilization of stripping membrane cascades intends to increase carbon dioxide removal for a fixed ethene loss, or reduce ethene losses for a fixed carbon dioxide removal. However, optimizations showed that the use of such membrane cascades is costlier than the use of single membrane modules or even the standalone absorption for the given application. This can be explained by the high cost of recompressing the permeate gas stream for recycle, or purging the gas stream containing ethene.

The diameters of the absorber and stripper do not change significantly between the standalone absorption and hybrid process with a single PI module for cases I and II. For cases III and IV the columns for the hybrid process have a smaller diameter as the total gas flow is significantly reduced by partially removing the CO_2 in the membrane section. Investment cost for the membrane material, which is to be replaced over time, is relatively small compared to the cost of the carbon steel casting. These are both relatively small if compared to the cost of the absorption and stripping columns with structured packing in stainless steel. Thus, the hybrid process also has the potential to offer a reduction in the capital investment costs.

Conclusions

MOSAIC is a convenient platform for documentation-based modeling of unit operations for subsequent export and integration into flowsheets. It allows the user to concentrate fully on the modeling activity, avoiding typographic mistakes and difficulties in model reuse, which are typical for coding in different environments. In this contribution, a gas separation membrane module is successfully exported from MOSAIC and integrated into Aspen for simulations and optimizations.

While very attractive for applications like natural gas sweetening or biogas enrichment, the utilization of gas permeation membranes for the OCM process is highly sensitive to product (ethene) losses. Hybrid processes employing a single PI membrane offer an economic advantage for OCM when CO_2 dilution is used in the reactor. Applying gas permeation membranes in this case leads to a significant reduction in the operational costs. For scenario IV, this reduction amounts to 103.39 US$/h. The use of CO_2 dilution is especially interesting for combining OCM with Dry Reforming of Methane as described in [18].

Outlook

A more effective and efficient approach for process optimization needs to be developed. The use of simplified models in an initial stage, especially in the absorption section, could ease the optimization and pave the way for the subsequent implementation of more rigorous models. Despite being a convenient platform for flowsheeting and preliminary cost analysis, Aspen Plus® is far from being a powerful optimization tool. The use of other tools needs to be considered for carrying a proper process optimization, notably simultaneous optimization, for the entire system. To this end, MOSAIC offers a built-in connection with the NEOS Server [19], which is a web-based service for optimization using state-of-the-art solvers.

Capital investment estimations need to be performed for the most promising scenarios in order to properly assess the advantages of the hybrid process. Cost of membrane units and membrane material replacement are yet to be estimated, and both negatively impact the hybrid process. More detailed economic analyses need to be performed, including the determination of the return on investment and payback times.

Finally, it is necessary to evaluate the operational and capital investment costs of the OCM process as a whole to assess its feasibility. Alternative unit operations like membrane reactors and pressure and temperature swing adsorption need to be considered in the OCM process superstructure. The integration of OCM with other processes as ethane cracking and methane dry reforming also need to be properly assessed.

Acknowledgement

Present work is developed with the support from CAPES, Coordination for the Improvement of Higher Education Personnel – Brazil (Grant Number BEX 11946/13-0). Financial support by the Cluster of Excellence "Unifying Concepts in Catalysis" (DFG EXC 314) is gratefully acknowledged.

Nomenclature

Latin letters

Ea	activation energy	kJ/mol
f	fugacity	Pa
h	membrane height	m
m^T	parameter in free-volume theory	K^{-1}
m^0	parameter in free-volume theory	Pa^{-1}
n	molar flow	kmol/s
z	membrane length	m
L	permeance	kmol/(m^2 s Pa)
R	universal gas constant	kJ/kmolK
T	temperature	K
y	molar fraction	kmol/kmol

Greek letters

σ	Lennard-Jones molecular diameter	Å

Superscripts

P	permeate
R	retentate
0	reference condition
PI	polyimid membrane
PEO	poly-(ethylene oxide) membrane

Subscripts

i	component index

References

[1] W. True, "Global ethylene capacity continues advance in 2011," *Oil & Gas Journal,* vol. 110, no. 7, 2012.

[2] A. T. Ashcroft, A. K. Cheetham, J. S. Foord, M. L. H. Green, C. P. Grey, A. J. Murrell and P. D. F. Vernon, "Selective oxidation of methane to synthesis gas using transition metal catalysts," *Nature,* vol. 344, pp. 319-321, 1990.

[3] Y. S. Su, J. Y. Ying and W. H. Green Jr., "Upper bound on the yield for oxidative coupling of methane," *Journal of Catalysis,* vol. 218, no. 2, p. 321–333, 2003.

[4] E. Esche, D. Müller, S. Song and G. Wozny, "Optimization during the process synthesis: enabling the oxidative coupling of methane by minimizing the energy required for the carbon dioxide removal," *Journal of Cleaner Production,* vol. 91, pp. 100-108, 2015.

[5] T. Brinkmann, C. Naderipour, J. Pohlmann, J. Wind, T. Wolff, E. Esche, D. Müller, G. Wozny and B. Hoting, "Pilot scale investigations of the removal of carbon dioxide from hydrocarbon gas streams using poly(ethyleneoxide)–poly(butyleneterephthalate) PolyActiveTM) thin film composite membranes," *Journal of Membrane Science,* vol. 489, p. 237–247, 2015.

[6] S. Kuntsche, H. Arellano-Garcia and G. Wozny, "MOSAIC, an environment for web-based modeling in the documentation level," *Computer Aided Chemical Engineering,* vol. 29, pp. 1140-1144, 2011.

[7] Aspen Technology, Inc., "Rate-Based Model of the CO2 Capture Process by MEA using Aspen Plus," Burlington, MA, USA, 2014.

[8] K. Ohlrogge und K. Ebert, Membranen: Grundlagen, Verfahren und industrielle Anwendungen, Weinheim, Germany: WILEY-VCH Verlag, 2006.

[9] J. J. Carroll and A. E. Mather, "A model for the solubility of light hydrocarbons in water and aqueous solutions of alkanolamines," *Chemical Engineering Science,* vol. 52, no. 4, pp. 545-552, 1997.

[10] B. Hanley and C.-C. Chen, "New Mass-Transfer Correlations for Packed Towers," *AIChE Journal,* vol. 58, no. 1, pp. 132-152, 2012.

[11] R. Taylor and R. Krishna, Multicomponent Mass Transfer, New York: Wiley, 1993.

[12] J. L. Bravo, J. A. Rocha and J. R. Fair, "A comprehensive model for the performance of columns containing structured packings," *I. ChemE. Symp,* vol. 129, no. A439, 1992.

[13] H. Hikita, S. Asai, H. Ishikawa and M. Honda, "The Kinetics of Reactions of Carbon Dioxide with Monoethanolamine, Diethanolamine, and Triethanolamine by a Rapid Mixing Method," *Chem. Eng. J.,* vol. 13, pp. 7-12, 1977.

[14] B. R. Pinsent, L. Pearson and F. J. W. Roughton, "The Kinetics of Combination of Carbon Dioxide with Hydroxide Ions," *Trans. Faraday Soc.,* vol. 52, pp. 1512-1520, 1956.

[15] E. Esche, H. Arellano-Garcia, G. Wozny and L. T. Biegler, "Optimal Operation of a Membrane Reactor Network," *Computer Aided Chemical Engineering,* vol. 31, p. 1321–1325, 2012.

[16] H. R. Godini, S. Jašo, S. Xiao, H. Arellano-Garcia, M. Omidkhah and G. Wozny, "Methane Oxidative Coupling: Synthesis of Membrane Reactor Networks," *Ind. Eng. Chem. Res.,* vol. 51, no. 22, pp. 7747-7761, 2012.

[17] A. Platts.com news & Analysis Feature, "www.platts.com," PLATTS McGraw Hill Financial, [Online]. Available: http://www.platts.com/news-feature/2014/petrochemicals/pgpi/ethylene. [Accessed 2015 07 07].

[18] H. R. Godini, S. Xiao, M. Kim, O. Görke, S. Song and G. Wozny, "Dual-membrane reactor for methane oxidative coupling and drymethane reforming: Reactor integration and process intensification," vol. 74, pp. 153-164, 2013.

[19] Wisconsin Institutes for Discovery at the University of Wisconsin in Madison, "NEOS Server," [Online]. Available: http://www.neos-server.org/neos/. [Accessed 17 09 2015].

434107 Membrane-Based Process for the Continuous Enzymatic Saccharification of Lignocellulosic Biomass

Abstract:

We are currently evaluating the feasibility of performing continuous enzymatic hydrolysis of lignocellulosic biomass to product sugars using a membrane-assisted reaction/separation process. The overarching technical goals are to continuously remove the sugars—this lowers product feedback inhibition—retain and recycle active enzyme, and continuously recover the co-product of lignin. Experimental data is being collected to support preliminary technoeconomic analysis of the speculative process designs versus the current state-of-the-art batch enzymatic hydrolysis. Prospective savings are anticipated due to higher time-averaged productivity from capital equipment investments, and lower enzyme usage. Our initial studies included characterizing cake-layer resistances for varying slurry solids loadings during crossflow filtration. These results highlighted the fact that the membrane's nominal permselective properties are less important than its module configuration and ease-of-cleaning because the cake formation controls the flux and fine tunes the selectivity. Thus, our current studies include two aspects. Firstly, we are coupling tubular ultrafiltration membranes to a continuous stirred-tank reactor and continuously removing the sugars and recycling the bulk of the water back to the reactor after concentrating them. Overall reaction kinetics and yield are being assessed and compared with classical batch hydrolysis results. Secondly, we are identifying sufficient protocols for in-situ membrane cleaning and duty cycling.

435012 Recovery of Nutrients from Swine Wastewater Using Ultrafiltration: Applications for Microalgae Cultivation in Photobioreactors

Abstract:

Algal biomass has been identified as a promising feedstock that could be used in a number of industrial applications, including biofuel production, aquaculture, and pharmaceutical production. Through the use of sunlight and CO_2, microalgae species are capable of producing biomass rich in lipids and carbohydrates, which can be extracted from the plant material and used in the production of commodities such as biofuels. Algae have been shown to have higher rates of productivity and higher lipid contents than traditional bioenergy crops, and do not require high-quality land for cultivation, and would not compete with current agricultural products for space.

While the high productivity and lipid content of microalgae make it a promising feedstock for a number of industrial applications, there are several challenges associated with their cultivation in large-scale production systems. The production of microalgae can require significant amounts of nitrogen and phosphorus fertilizer. In order to make the production of microalgae economical, it is important that low cost sources of nitrogen and phosphorus be available to producers. Considering the relatively high nitrogen content of microalgae (4 to 8 percent), the nitrogen inputs required for microalgae cultivation can be significant. In addition, the synthesis of nitrogen fertilizer produces around 2 kg of CO_2 kg^{-1}. The additional carbon production, if added to the overall lifecycle, undermines the favorable carbon balance for microalgae feedstocks.

Membrane separations technology constitutes an alternative method for the removal of biological contaminants from the nutrient-rich wastewater prior to use in microalgae cultivation. A membrane consists of a thin film that separates two phases, and can be used in liquid-liquid separations. Microfiltration and ultrafiltration involve the use of porous membranes to remove micro- or macro-particles from a solution. In this study we used ultrafiltration membranes for removing contaminants from swine wastewater for use in algal growth. Through several trials we have found that the algal wastewater is completely contaminant free and the growth rate of the algae is greater than optimal conditions.

433601 Use of Forward Osmosis in Treatment of Hyper-Saline Produced Water

Abstract:

Oilfield produced water forms the largest waste stream in the production of petroleum. The quality of produced water varies greatly, with some waters being close to potable while others are oily or extremely saline. Hyper-saline produced water is an exceptionally challenging problem as there are few treatment options available. Desalination of such hyper-saline water is extraordinarily expensive because we are limited to evaporative processes which have enormous energy requirements. Conventional desalination technologies, like reverse osmosis are unusable due to the high osmotic pressure which require exceptionally high hydraulic pressure to overcome.

Forward osmosis circumvents this issue by using osmotic pressure instead of hydraulic pressure. Using a high concentration draw solution, osmotic potential can force osmotic flow from even the highest salinity solutions. The draw solution can then be subsequently regenerated (or used) by another process chosen specifically for that draw solute. While forward osmosis will not use less energy than other technologies, it does have the capability of using lower quality energy (low temperature thermal) or operating with other advantages (low fouling propensity, for instance). FO achieves this while relying simply on an osmotic potential gradient generated by a concentrated draw solution across a semi-permeable membrane. In this work, the feasibility of using FO in treatment of produced water with TDS of about 240,000 ppm was studied. Both an asymmetric cellulose triacetate (CTA) and thin film composite (TFC) forward osmosis (FO) membrane from Hydration Technologies Innovation (HTI) were used for this study. Synthetic produced water was used as a feed solution to simulate the composition of the oilfield produced water. We evaluated the individual effect of the different salts in produced water (NaCl, $CaCl_2$, $MgCl_2$, $MgSO_4$, $NaHCO_3$, and $FeCl_3$) on the flux performance. Two types of draw solutions were used in this study: 6M ammonia-carbon dioxide draw solution and 4.8M magnesium chloride draw solution. Results show that although NH_3-CO_2 has the advantage of being economically separated and recycled to the FO process, it has high reverse solute flux across the membrane. This reverse solute flux can also have unforeseen effects such as the formation of low solubility salts that cause scale or a pH change (in this case, a rise), which can also impact the solubility of salts. While ammonia is a key solute that crosses the membrane, carbonate can also precipitate readily if it diffuses. The use of $MgCl_2$ showed promise as it gave higher water flux with less scaling, although the recovery method (membrane distillation or evaporation) would have greater energy demands and higher cost.

425116 Membrane Distillation for Separating Water from Ionic Liquid Solutions

Abstract:

Ionic liquids, salts molten at temperatures below 100 °C, have exciting potential as "green" recyclable solvents for industrial processes. Ionic liquids have been found that can dissolve the lignin from biomass at relatively low temperatures. The pretreated biomass is then ready for hydrolysis and fermentation to produce biofuels. Water is added to the ionic liquid solution to surround the ions so that the dissolved lignin precipitates out. This water must be removed prior to recycling the ionic liquid for pretreatment. Two ionic liquids known to be effective in biomass pretreatment are 1-ethyl-3-methylimidazolium acetate and 1-ethyl-3-methylimidazolium formate. These were diluted with water and their conductivities and densities measured at varying concentrations. Membrane distillation of the ionic liquid – water mixtures was performed using two different well-characterized PTFE membranes. The feed was maintained at 55 °C and the distillate at 20 °C. The feed and distillate were circulated counter-currently on opposite sides of the membrane at 1.5 L/min. Water flux obtained was ~10 $L/m^2 \cdot h$. Concentration of IL in the feed concentrate and distillate was determined by conductivity and density measurements. The membrane distillation process provided a concentration factor of 10 (from 5% IL to 50% IL) with near 100% rejection; as no ionic liquid was detected in the final distillate. Membrane distillation is a promising technique for re-concentrating ionic liquids for recycling.

433013 Deployment of Explicit Model Predictive Controller Onboard a Mini Fuel Cell Vehicle

Abstract:

Deployment of Explicit Model Predictive Controller Onboard a Mini Fuel Cell Vehicle

Amit M. Manthanwar† and Efstratios N. Pistikopoulos‡

†Department of Chemical Engineering, Imperial College London, England, UK

‡Artie McFerrin Department of Chemical Engineering, Texas A&M University, USA

Abstract: 07 - TRANSPORT AND ENERGY PROCESSES

Targeted Session: 07000 Recent Advances in Fuel Cell and Battery Technologies

Fuel cells are electrochemical devices that convert hydrogen or hydrogen rich fuel directly into electricity. Fuel cells are more efficient compared to other combustion based technologies and offer a wide range of tangible environmental, economic and operational benefits. However, control is critical to the robust operation of the system—ultimately essential for maintaining uninterruptable power demand under load variations, [1]. Furthermore, fuel cell system integration to automotive powertrain brings additional challenges affecting durability and performance of the overall system. The majority of fuel cell system failures and forced outages (~90%) are due to lack of system integration, [2].

The real time control of fuel cell vehicle using a well established technology of model predictive control requires an accurate mathematical model of the system. The dynamic phenomenon taking place inside fuel cell involves complex interactions of mass transport, energy transport and electrochemical kinetics. This complexity further increases when the fuel cell dynamics are coupled with the vehicle transmission dynamics. These uncertain dynamics determine the overall operational performance of the vehicle. In addition, the real time control requires computing machinery to run the optimisation algorithms. This compounds the manufacturing and operational costs. In order to resolve these issues, we present (a) an experimentally validated dynamic model of the polymer electrolyte membrane fuel cell; (b) an experimentally validated model of DC motors used in a vehicle powertrain; and (c) generic control framework with deployment strategy of explicit model predictive controller, [3], for the efficient operation of the fuel cell vehicle propulsion system.

This work presents a prototype fuel cell vehicle with step by step procedure to deploy explicit MPC on a fast 'mbed' platform based on a high performance ARMr CortexTM M3 microcontroller. Our analysis of the experimentally achieved driving cycle results demonstrate that significant reduction in operational costs can be achieved while simultaneously improving the overall performance the fuel cell vehicle. Such modelling and control framework is critical to effective operation as well as preventive maintenance of the fuel cell vehicle. Thus, we conclude that proposed "MPC-on-a-chip" strategy is the best available option for making smarter decisions in operating fuel cell vehicles reliability and further improving their competitive advantage.

References

[1] C. Ziogou, E. N. Pistikopoulos, M. C. Georgiadis, S. Voutetakis, and S. Papadopoulou, "Empowering the performance of advanced nmpc by multiparametric programming–an application to a pem fuel cell system," Ind. Eng. Chem. Res., vol. 52, pp. 4863–4873, Mar. 2013.

[2] K. Wipke, S. Sprik, J. Kurtz, T. Ramsden, C. Ainscough, and G. Saur, "Final results from u.s. fcev learning demonstration," in EVS26 International Battery, Hybrid and Fuel Cell Electric Vehicle Symposium, (Los Angeles, California), May 6–9 2012.

[3] E. N. Pistikopoulos, "From multi-parametric programming theory to mpc-on-a-chip multi-scale systems applications," Computers and Chemical Engineering, no. 0, 2012.

434826 Combined Modeling and Experimental Analysis of Microscale Fuel Cells

Abstract:

Combined Modeling and Experimental Analysis of Microscale Fuel Cells

Adam S. Hollinger, Daniel G. Doleiden

Department of Mechanical Engineering

Penn State Erie, The Behrend College, 5101 Jordan Road, Erie, PA 16563, USA

Section: Transport and Energy Processes (07)

Session: 07000 Recent Advances in Fuel Cell and Battery Technologies

With the rapid advancement of consumer mobile technology, an accompanying demand for energy-dense portable power sources has given rise to increased development of alternatives to traditional rechargeable battery technologies (i.e., lithium-ion) [1]. Microscale direct methanol fuel cells (DMFCs) demonstrate potential as such an alternative, enabling instantaneous recharging via fuel solution cartridges and higher energy density than lithium-ion rechargeable batteries [1].

When optimizing the design of DMFCs for portable power applications, laboratory testing involving the production of fuel solutions and iterative fabrication of various components is typically required [2]. Here, we present a UNIX-based mathematical model of flow and diffusion through the flow chamber of a DMFC as a means of predicting the effects of various parameters on cell performance. This model will aid design optimization by reducing the amount of time and materials required for experimental benchmarking.

After input parameters are defined, the model provides expeditious 2D and 3D graphical simulation of current density, fuel molarity, and hydrodynamic flow as functions of position in the flow chamber. In addition, polarization curves can be generated from model output for comparison to experimental data. Such comparisons enable internal modifications of the model for increased accuracy.

A positive correlation between model predictions and experimental data for catalyst deposition width studies at low to medium power density has already been observed; evaluation of the model's ability to predict the effects of varying fuel molarity, electrolyte molarity, flow chamber geometry, and fuel flow rate are ongoing.

[1] E. Kjeang, N. Djilali, D. Sinton, Journal of Power Sources, 186 (2009) 353-369.

[2] A.S. Hollinger, P.J.A. Kenis, Journal of Power Sources, 240 (2013) 486-493.

428356 Metal-Catalyst-Free Fuel Cells with Nanocarbon As Catalysts

Abstract:

Fuel cells are considered as a potential solution to sustainable energy, however their wide spread commercialization have long been hindered by limited reserve and relatively high cost of metal catalysts. Although significant progress has been made to explore non-noble metal catalysts, it remains challenging to replace metal catalysts at the anode side of low temperature fuel cells by non-metal catalysts, thus achieving total removal of the metal catalysts from a fuel cell. We present that 3D graphene prepared by reduction of carbon monoxide with lithium oxide, can serve as as a carbon-only catalyst for fuel cell, electrochemically catalyzing carbohydrazide oxidation reaction efficiently. Muti-walled CNT also outperforms carbon black and 2D graphene in catalyzing carbohydrazide oxidation reaction. Prototypes of anode metal free and completely metal free anion exchange membrane fuel cells with 3D graphene anode catalyst generate peak power density of 75.1 mW cm^{-2} and 24.9 mW cm^{-2}, respectively. In average, 4.9 electrons are electrochemically extracted from one carbohydrazide molecule, indicating the existence of C-N bond cleavage, which is a key factor contributing to high fuel utilization efficiency.

Acid Leaching Model for HT-PEMFCs – Stiction & Dilution Approach

Srinivasan Raman[*], Raghunathan Rengaswamy[*,+]

[*] Department of Chemical Engineering, Texas Tech University, USA
[+] Department of Chemical Engineering, IIT – Madras, India

E-mail: srinivasan.raman@ttu.edu, raghu.rengasamy@ttu.edu

Presenter: T. Hemanth Kumar

1

SECTIONS

1. Introduction
2. Acid leaching model development
3. Results

2

INTRODUCTION

HT-PEMFCs :: Advantages :: Issues

3

High Temperature PEM Fuel Cells

- Operating temperature – above 100 °C
  - Nafion cannot be used
- Proton Exchange Medium – PBI with H_3PO_4
- Advantageous compared to low temperature PEM-FCs

4

Advantages of HTPEM-FCs

- Improved kinetics of reactions at the electrode.
- Simplified water management – gaseous state water.
- Reactants and products in the same phase.
- Simplified cooling system.
- Recovery of waste heat for practical use.
- Increased CO tolerance.

5

Acid Leaching

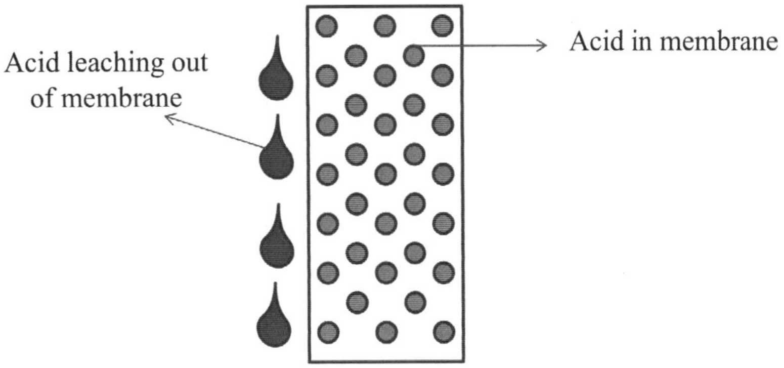

- Irreversible loss of H_3PO_4 from PBI membrane
- Loss of fuel cell performance
- Acid leaching to be minimized – improved performance
- Simple first principles model needed

6

ACID LEACHING MODEL DEVELOPMENT

Conceptualization :: Estimation :: Effects

7

Model Assumptions

- Temperature of operation – 120 °C to 155 °C
- Water is in gaseous phase and phosphoric acid doesn't evaporate
- Gaseous phase reactants and products
- No gas crossover
- Ideal gas behavior
- Change in membrane dimensions during operation is neglected

8

Conceptualization

- Acid held in pores – Capillary forces
- Water formed at cathode
- Diffusion towards anode
- Dilution of acid
- Viscosity reduces due to diffusion of water
- Flows out at threshold

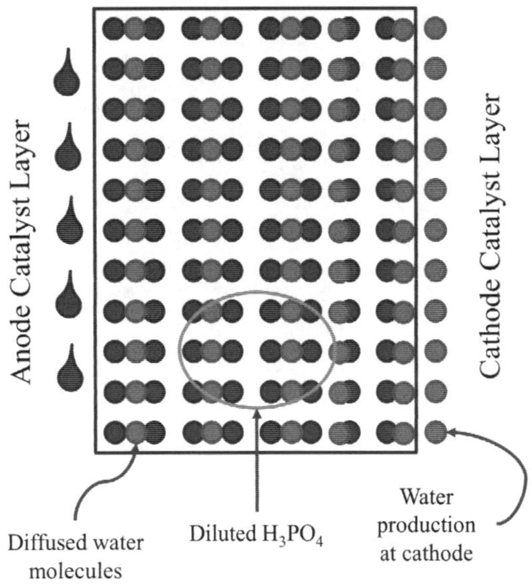

9

Diffusivity Prediction

- Method by Akgerman et al. – diffusion of solute in solvent
- Predicted diffusivity of water in pure phosphoric acid:
  - 3.2567×10^{-8} cm²/s.
- Effective diffusivity – tortuous path and porosity of PBI considered

$$D_{eff} = \frac{\epsilon D}{\tau}$$

$$D_{eff} = \frac{D}{6} = 5.4278 \times 10^{-9}\, cm^2/s$$

10

PBI Membrane Assumptions

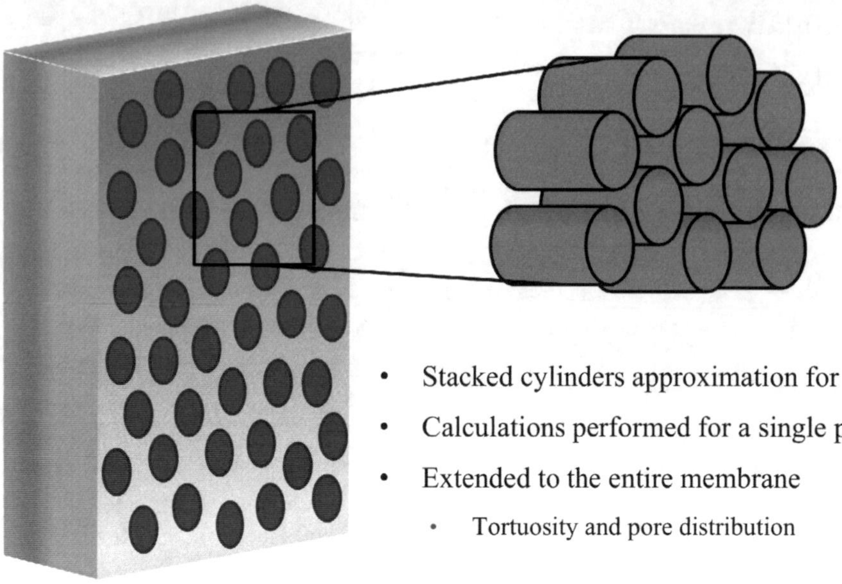

- Stacked cylinders approximation for pores
- Calculations performed for a single pore
- Extended to the entire membrane
  - Tortuosity and pore distribution

11

Capillary Action

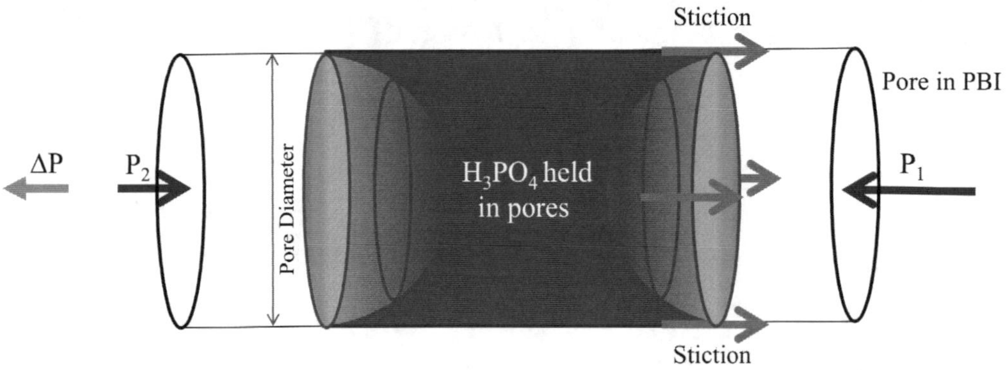

- P_1 – Cathode side pressure
- P_2 – Anode side pressure
- $P_2 < P_1$ since more cathode gases are assumed to be supplied for kinetic reasons, giving rise to ΔP
- Stiction holds the H_3PO_4 in the pores of the membrane, when the diluted acid gets pushed by the pressure forces

12

Descriptive Equations

- Surface forces: $F_{surface} = \Gamma . 2\pi r$
- Pressure forces: $F_{pressure} = \Delta P . \pi r^2$
- At equilibrium: $F_{surface} = F_{pressure}$
- Surface Tension: $\Gamma = \frac{\Delta P . r}{2}$

- Surface tension – Viscosity relationship: $\Gamma = Ae^{-\frac{B}{\eta_{absolute}}}$

$[A, B - f(T)]$

$$\Gamma_{H_3PO_4} = A.e^{-\frac{B}{\eta_{yield}}} = \frac{\Delta P . r}{2}$$

$$\boxed{\eta_{yield} = -\frac{B}{\ln\left(\frac{\Delta P . r}{2.A}\right)}}$$

13

Limiting Factor

$$if\ \eta_{absolute,eff} > \eta_{yield} \rightarrow Pressure\ forces\ balanced\ by\ stiction$$

$$if\ \eta_{absolute,eff} < \eta_{yield} \rightarrow Acid\ leaching$$

$$Amount\ of\ acid\ leached = \boxed{Volumetric\ flow\ rate\ of\ acid\ from\ pores} \times time$$

14

- Hagen – Poiseuille's equation: $\frac{dV}{dt} = \frac{\pi}{8} \frac{r^4}{\eta_{absolute,eff}} \frac{\Delta P}{L}$

$$if\ \eta_{absolute,eff} > \eta_{yield}\ , V_{acid\ per\ pore} = 0$$

$$if\ \eta_{absolute,eff} < \eta_{yield}\ , V_{acid\ per\ pore} = t \times \frac{\pi}{8} \frac{r^4}{\eta_{absolute,eff}} \frac{(\text{Driving Force})}{L}$$

$$Driving\ Force = Pressure\ forces - Stiction$$

$$Number\ of\ pores\ (n_{pores}) = Area\ of\ membrane \times Number\ of\ pores\ per\ unit\ area\ of\ the\ membrane$$

15

Connection of Acid Leaching with Fuel Cell Performance

- Assumed form of effective proton conductivity: $k_{eff} = \sigma_{eff,H_3PO_4} + \sigma_{eff,H_2O}$
- Assumed functional form for proton conductivity terms:

$$\sigma_{eff,H_3PO_4} \propto (n_{total}, \sigma_{H_3PO_4}, x_{H_3PO_4})$$

$$\sigma_{eff,H_3PO_4} = K_{H_3PO_4} \times n_{total} \times \sigma_{H_3PO_4} \times x_{H_3PO_4}$$

Similarly,

$$\sigma_{eff,H_2O} = K_{H_2O} \times n_{total} \times \sigma_{H_2O} \times x_{H_2O}$$

- Effect of acid leaching: decrease in moles of acid & water ➔ decrease in k_{eff}

16

Fuel Cell Voltages

Cell Voltage
= Thermodynamic potential – Ohmic losses – Overpotentials at cathode and anode

$$V_{cell} = E_{therm} - \eta_{ohmic} - \eta_{act}^{a} - \eta_{act}^{c}$$

$$\eta_{ohmic} = j_{cell}\left[\left(\frac{\delta^{m}}{k}\right) + \left(\frac{\delta^{a}}{\sigma^{a}}\right) + \left(\frac{\delta^{c}}{\sigma^{c}}\right)\right]$$

17

RESULTS

Threshold :: Response :: Consequence

18

Viscosity change of electrolyte in the membrane with fuel cell operating time due to water diffusion

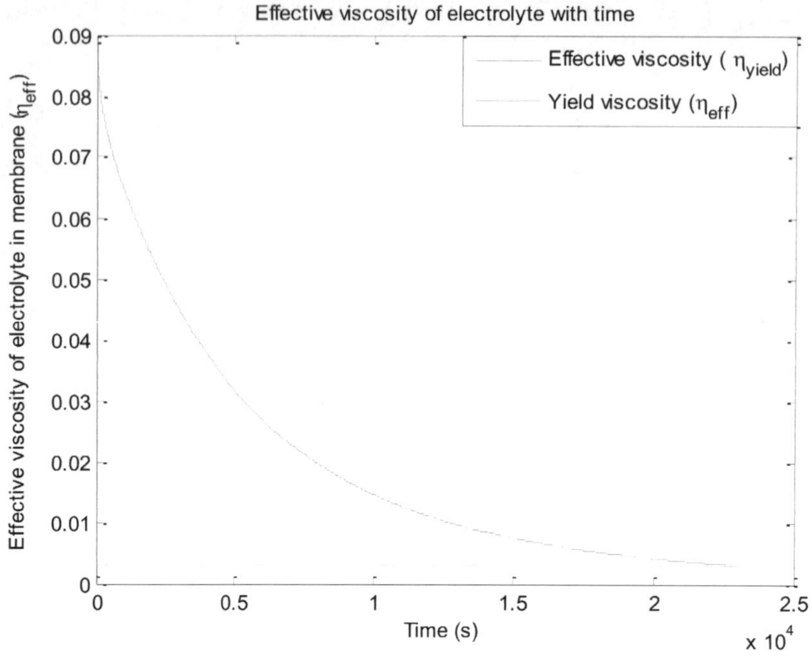

19

Plot of proton conductivity of membrane against the loss of electrolyte due to acid leaching

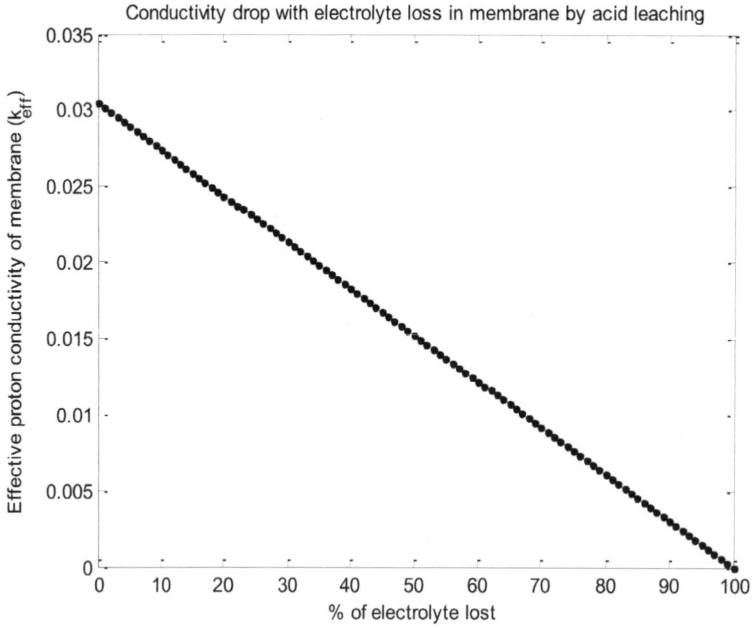

20

Performance curve for HTPEMFC at different values of proton conductivities

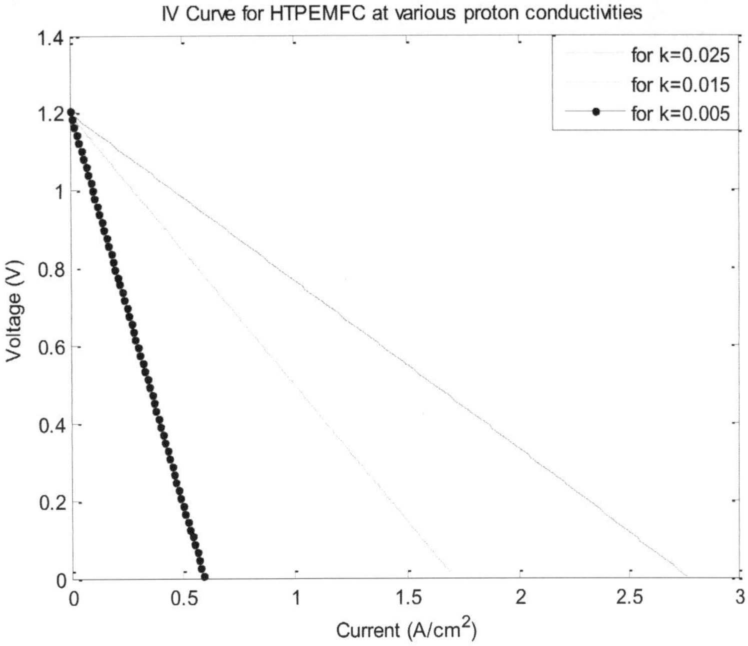

21

Conclusions

- Developed a simple model for acid leaching in HTPEMFCs
- Can be used to dynamically predict acid leaching
- Experimental data has been generated

Future Work

- Adding more mechanisms to this simple model
- To define operational strategies to minimize acid leaching
- Verification of results with available experimental data

22

Thank You

23

REFERENCES

1. Larminie J, Dicks A, McDonald MS. Fuel cell systems explained. Wiley New York; 2003.
2. Zhang J. PEM fuel cell electrocatalysts and catalyst layers: fundamentals and applications. Springer; 2008.
3. Marx D. Proton transfer 200 years after von Grotthuss: insights from ab initio simulations. Chemphyschem [Internet]. 2006 Sep 11 [cited 2014 Jun 4];7(9):1848–70. Available from: http://www.ncbi.nlm.nih.gov/pubmed/16929553
4. Shao Y, Yin G, Gao Y. Understanding and approaches for the durability issues of Pt-based catalysts for PEM fuel cell. J Power Sources [Internet]. 2007 Sep [cited 2014 Feb 27];171(2):558–66. Available from: http://linkinghub.elsevier.com/retrieve/pii/S0378775307014814
5. Borup R, Meyers J, Pivovar B, Kim YS, Mukundan R, Garland N, et al. Scientific aspects of polymer electrolyte fuel cell durability and degradation. Chem Rev [Internet]. 2007 Oct;107(10):3904–51. Available from: http://www.ncbi.nlm.nih.gov/pubmed/17850115
6. Chandan A, Hattenberger M, El-kharouf A, Du S, Dhir A, Self V, et al. High temperature (HT) polymer electrolyte membrane fuel cells (PEMFC) – A review. J Power Sources [Internet]. Elsevier B.V; 2013 Jun [cited 2013 May 21];231:264–78. Available from: http://linkinghub.elsevier.com/retrieve/pii/S0378775312018113
7. Zhang J, Xie Z, Zhang J, Tang Y, Song C, Navessin T, et al. High temperature PEM fuel cells. J Power Sources [Internet]. 2006 Oct [cited 2014 May 27];160(2):872–91. Available from: http://linkinghub.elsevier.com/retrieve/pii/S0378775306010299
8. Lin HL, Hsieh YS, Chiu CW, Yu TL, Chen LC. Durability and stability test of proton exchange membrane fuel cells prepared from polybenzimidazole/poly(tetrafluoro ethylene) composite membrane. J Power Sources [Internet]. 2009 Aug [cited 2014 May 27];193(1):170–4. Available from: http://linkinghub.elsevier.com/retrieve/pii/S0378775309000664
9. Darling RM, Meyers JP. Kinetic Model of Platinum Dissolution in PEMFCs. J Electrochem Soc [Internet]. 2003 Nov [cited 2013 Aug 22];150(11):A1523–A1527. Available from: http://jes.ecsdl.org/content/150/11/A1523
10. Parthasarathy P, Virkar A V. Electrochemical Ostwald ripening of Pt and Ag catalysts supported on carbon. J Power Sources [Internet]. Elsevier B.V; 2013 Jul [cited 2013 May 29];234:82–90. Available from: http://linkinghub.elsevier.com/retrieve/pii/S0378775313001729

24

11. Liu G, Zhang H, Hu J, Zhai Y, Xu D, Shao Z. Studies of performance degradation of a high temperature PEMFC based on H3PO4-doped PBI. J Power Sources [Internet]. 2006 Nov [cited 2014 May 27];162(1):547–52. Available from: http://linkinghub.elsevier.com/retrieve/pii/S0378775306012468

12. Jerkiewicz G, Vatankhah G, Lessard J, Soriaga MP, Park Y-S. Surface-oxide growth at platinum electrodes in aqueous H2SO4. Electrochim Acta [Internet]. 2004 Apr [cited 2014 Feb 8];49(9-10):1451–9. Available from: http://linkinghub.elsevier.com/retrieve/pii/S0013468603009277

13. Zhao Z, Castanheira L, Dubau L, Berthomé G, Crisci a., Maillard F. Carbon corrosion and platinum nanoparticles ripening under open circuit potential conditions. J Power Sources [Internet]. Elsevier B.V; 2013 May [cited 2013 May 29];230:236–43. Available from: http://linkinghub.elsevier.com/retrieve/pii/S037877531201899X

14. De Beer C, Barendse P, Pillay P, Bullecks B, Rengaswamy R. Degradation of high temperature PEM fuel cells and the impact on electrical performance. 2013 IEEE Int Conf Ind Technol [Internet]. Ieee; 2013 Feb;690–4. Available from: http://ieeexplore.ieee.org/lpdocs/epic03/wrapper.htm?arnumber=6505755

15. Boaventura M, Sousa JM, Mendes a. A dynamic model for high temperature polymer electrolyte membrane fuel cells. Int J Hydrogen Energy [Internet]. Elsevier Ltd; 2011 Aug [cited 2014 Jun 4];36(16):9842–54. Available from: http://linkinghub.elsevier.com/retrieve/pii/S0360319911011177

16. Cutlip MB, Shacham M. Problem solving in chemical engineering with numerical methods. Prentice Hall PTR Englewood Cliffs; 1999.

17. Wainright JS, Wang J, Weng D, Savinell RF, Litt M. Acid-Doped Polybenzimidazoles : A New Polymer Electrolyte. 1995;142(7):121–3.

18. He R. Proton conductivity of phosphoric acid doped polybenzimidazole and its composites with inorganic proton conductors. J Memb Sci [Internet]. 2003 Dec 1 [cited 2014 May 27];226(1-2):169–84. Available from: http://linkinghub.elsevier.com/retrieve/pii/S0376738803004216

19. He R, Li Q, Bach a, Jensen J, Bjerrum N. Physicochemical properties of phosphoric acid doped polybenzimidazole membranes for fuel cells. J Memb Sci [Internet]. 2006 Jun 1 [cited 2014 May 27];277(1-2):38–45. Available from: http://linkinghub.elsevier.com/retrieve/pii/S0376738805007337

20. Bouchet R, Miller S, Duclot M, Souquet J. A thermodynamic approach to proton conductivity in acid-doped polybenzimidazole. Solid State Ionics [Internet]. 2001 Dec 1;145(1-4):69–78. Available from: http://linkinghub.elsevier.com/retrieve/pii/S0167273801009158

21. Geankoplis C. Transport processes and separation process principles (includes unit operations). Prentice Hall Press; 2003.

22. Akgerman A, Gainer JL. Predicting Gas-liquid diffusivities. J Chem Eng Data. 1972;17(3):372–7.

23. Pribil AB, Hofer TS, Randolf BR, Rode BM. Structure and Dynamics of Phosphate Ion in Aqueous Solution : An Ab Initio QMCF MD Study. 2008;

24. Peter A, Julio de P. Atkins' Physical Chemistry. 8th ed. W. H. Freeman; 2006.

25. Cussler EL. Diffusion: mass transfer in fluid systems. Cambridge university press; 2009.

26. Halliday D, Resnick R, Walker J. Fundamentals of physics extended. John Wiley & Sons; 2010.

27. Laboratories BT, Hill M. Surface Tension-Viscosity Relationship for liquids. (7):524–5.

28. Phosphoric Acid Properties [Internet]. INNOPHOS PHOSPHORIC ACID. [cited 2014 Jun 4]. Available from: http://www.innophos.com/__sitedocs/journal_articles/innophos-phosphoric-acid-outline.pdf

29. Perry RH, Green DW, Maloney JO. Perry's chemical engineer's handbook. Perry's chemical engineer's handbook. McGraw-Hill Book; 1984.

30. Wang P, Anderko A, Young RD. Modeling Surface Tension of Concentrated and Mixed-Solvent Electrolyte Systems. Ind Eng Chem Res [Internet]. 2011 Apr 6;50(7):4086–98. Available from: http://pubs.acs.org/doi/abs/10.1021/ie101915n

31. Astolfi A. An introduction to Optimization. 2006;(September 2002). Available from: http://www3.imperial.ac.uk/pls/portallive/docs/1/7288263.PDF

26

408675 One Dimensional Theoretical Model for Water Transport in PEM Fuel Cell

Abstract:

Model plays an important role in fuel cell and offer an innovative alternative to current power sources with potential two achieve higher efficiencies with renewable fuels with minimal environmental impact. Model predictions of the performance of the fuel cell are comparison with experimental and data published by Hung et al. (2007) and compared with recent values published by Liu et al.(2007). The mass and heat transport in the model are steady state and one – dimensional. The development of model consists of various plates and anodic and cathodic reactions takes place at the anode side and cathode side and it relies on various assumptions. The model is consisting of mass transfer model and heat transfer model. The operating conditions considered are various cathode and anode relative humidity (RH) values at 80° C and 2atm. The heat generated in the cell is partly lost to the ambient and partly removed by the cathode and anode fuel streams.In this model only activation and ohmic losses were considered .The numerical tools used to implement the model were MATLAB and EXCEL. A one dimensional water and thermal management model considering the effects of water transport across the membrane, activation, ohmic and concentration overpotentials, pressure drops and current density distribution along the channel of a PEM (proton exchange membrane).Model predictions were successfully compared to experimental and theoretical I-V polarization curves presented by Hung et al.(2007) and JU et al.(2005).Humidified cathode and specially humidified anode streams are needed to avoid the membrane dehydration, particularly at high current densities and it provide suitable operating ranges adequate to different applications for low humidity operation.

417401 Highly Active and Durable Extended Surface Oxygen Reduction Electrocatalysts

Abstract:

The amount of platinum (Pt) in the catalyst layer accounts for a significant portion of fuel cell cost and limits the commercial deployment of proton exchange membrane fuel cells.[1] Extended surface nanomaterials have been developed as electrocatalysts in the oxygen reduction reaction for use in fuel cells. Extended surfaces offer key advantages to nanoparticles, including an order of magnitude higher specific activity, long range conductivity, and long term durability, but are traditionally limited by low surface area.[2] Catalyst development has included a variety of approaches, but recently has focused on galvanic displacement.

Spontaneous galvanic displacement occurs when a less noble metal template contacts a more noble metal cation and combines aspects of corrosion and electrodeposition. Catalysts formed by spontaneous galvanic displacement are ideally situated, being able to take advantage of the specific activities generally associated with the catalyst type while significantly improving upon the surface area.[3] Recent developments in Pt-nickel (Ni) nanowires have produced materials with surface areas in excess of 90 m^2 g_{Pt} ^{1}.[4]

Post-synthesis processing has been used to alloy the Pt-rich and Ni-rich zones, improving activity for oxygen reduction. Exposure to a variety of acid types and concentrations has been used to remove increasing amounts of Ni, creating nanostructures with high Pt surface areas and an array of compositions. Oxidative treatments have also been used to improve catalyst durability in rotating disk electrode (RDE) accelerated stress tests (30,000 potential cycles, 0.6 1.0 V). By optimizing these techniques, extended surface electrocatalysts have been created with oxygen reduction mass activities 7 times greater than Pt nanoparticles and 5 times greater than the U.S. Department of Energy membrane electrode assembly (MEA) target in RDE half-cells. In accelerated stress tests, these materials lose less than 3% activity and less than 0.5% of their mass due to electrochemical dissolution.

Remaining barriers, including synthesis scalability and MEA fabrication, are being addressed. Recent improvements in MEA performance suggest that the promise of extended surface catalysts in half-cells can be realized in the device.

[1] D. Papageorgopoulos, in: U.S. Department of Energy (Ed.), http://www.hydrogen.energy.gov/pdfs/review14/fc000_papageorgopoulos_2014_o.pdf, 2014.

[2] M. Debe, in: U.S. Department of Energy (Ed.), http://www.hydrogen.energy.gov/pdfs/review09/fc_17_debe.pdf, 2009.

[3] S.M. Alia, Y.S. Yan, B.S. Pivovar, Catalysis Science & Technology, 4 (2014) 3589-3600.

[4] S.M. Alia, B.A. Larsen, S. Pylypenko, D.A. Cullen, D.R. Diercks, K.C. Neyerlin, S.S. Kocha, B.S. Pivovar, ACS Catalysis, 4 (2014) 1114-1119.

424080 $LiNi_{0.5}Mn_{1.5}O_4$ Spinel Material for Application As Solid Oxide Fuel Cell Cathode

Abstract:

$LiNi_{0.5}Mn_{1.5}O_4$ (LNM) spinel is a commercial positive electrode material for lithium batteries. This material is attractive as oxygen vacancies are formed in this spinel material, which is an essential requirement for SOFC materials. We found that the LNM spinel showed rapid oxygen release and uptake rates and that a large amount of oxygen vacancies was formed Thus, here we attempt to utilize this spinel as a cathode material for SOFC.

All the cathode material requirements have been investigated. The spinel showed excellent chemical stability from 1000 to 1400 °C in air, while it underwent reduction in H_2 at 800 °C. No interaction between LNM and yttria-stabilized zirconia (YSZ) electrolyte was observed by XRD and EDS characterizations. The total conductivity of LNM was about 0.2-5 S cm^{-1} at the temperature of 400-800 °C by 4-probe DC test. The LNM was spin-coated on both sides of an YSZ electrolyte support to form a symmetry cell for electrochemical performance investigation with gold paste as current collector. The resistance of the LNM cathode was 0.77 Ω cm^2 at 800 °C according to impedance spectroscopy measurements (EIS), better than traditional $La_{1-x}Sr_xMnO_3$ perovskite cathode material. A long term study of cathode impedance was also conducted. It indicated that the cathode material can uptake oxygen from the atmosphere gradually at temperatures lower than 700 °C, leading to a decrease of the oxygen vacancy amount and an increase of cathode resistance with time.

Further studies are focusing on investigating the oxygen exchange property of LNM spinel from electrochemical aspects, in order to get a better understanding of the nature between the oxygen vacancy and the electrochemical properties of material.

420966 Fuel Flexible Solid Oxide Fuel Cells with High Power Density: Potential for Future Hybrid Electric Vehicles

Abstract:

Our research describes the performance of a solid oxide fuel cell (SOFC) fueled by directly feeding either premium gasoline or biodiesel to the anode without using external reforming. The novel component of the fuel cell that enables such operation is the mixed conductivity MoO_2-based anode .This unusual ceramic material exhibits excellent electronic/ionic conductivity, and this behavior is believed to lead to high catalytic activity of MoO_2 towards partial oxidation of various hydrocarbons. In addition, MoO_2 shows high resistance to coke formation, which allows longer and more stable operation. Using this anode with premium gasoline as the fuel, a fuel cell demonstrating a power density > 3.0 W cm^{-2} at 0.6 V was successfully fabricated. Over a 24 h period of operation, the open cell voltage remained stable at ~0.9 V. At the cell voltage of 0.6 V, its current density dropped over the first 7 h to a value of ~3.0 A cm^{-2}, where it stayed for the remaining 17 h of the test with a minor fluctuation. Power density of ~2.0 W cm^{-2} at 0.6 V was still measured after 24 h on stream with a continuous feed of gasoline. Scanning electron microscopy (SEM) examination of the anode surface pre- and post-testing showed no evidence of coking, which hints at the reason for the observed stability under the harsh cell operating conditions. The implication of this preliminary study is that an SOFC using a MoO_2-based anode has potential for generating electrical power from both gasoline and biodiesel for future hybrid electric vehicles.

High Throughput Approach to Accelerate Electrolyte discovery for batteries

Nav Nidhi Rajput Postdoc[1], Xiaohui Qu Postdoc[1] and Kristin A. Persson Staff Scientist[1]

[1]*Electrochemical Technologies Group, Lawrence Berkeley National Laboratory, Berkeley, California, 94702, USA*

ABSTRACT

Increasing demand of high energy density and high capacity batteries require development beyond Li-ion technology and innovations in electrodes and electrolytes, alike. An approach to achieve energy density higher than the existing Li-ion batteries is to replace monovalent Li ion by multivalent ions such as Mg, Ca. The ability of Mg batteries to provide much higher volumetric capacity, particularly on the anode side where the Mg metal can theoretically provide 3833 mA h/cc as compared to the Li counterpart graphite (□ 800 mA h/cc) at a lower cost makes the technology an attractive candidate for future batteries. However, development and commercialization of Mg batteries require not only improved electrode discovery and development but also novel electrolytes, which are compatible with the Mg metal as conventional electrolytes fail to penetrate the Mg metal passivation layer. Thus a fundamental understanding of molecular level properties of these electrolytes is required to improve the electrochemical stability and the charge transfer properties. An automatic High-throughput infrastructure has been constructed for the electrolyte genome project supported by the US Joint Center for Energy Storage Research (JCESR)[2]. In this work, we present classical molecular dynamics simulations coupled with *ab initio* calculations for Mg salts in various solvents. We uncover a novel effect between concentration dependent ion pair formation and anion stability at reducing potential, e.g., at the metal anode[1]. We elucidate systematic correlations between molecular level interactions and composite electrolyte properties, such as electrochemical stability, solvation structure, and dynamics. We find that Mg electrolytes are highly prone to ion pair formation, even at modest concentrations, for a wide range of solvents with different dielectric constants, which have implications for dynamics as well as charge transfer. Specifically, we observe that, at Mg metal potentials, the ion pair undergoes partial reduction at the Mg cation center ($Mg^{2+} \rightarrow Mg^{+}$), which competes with the charge transfer mechanism and can activate the anion to render it susceptible to decomposition. Specifically, $TFSI^-$ exhibits a significant bond weakening while paired with the transient, partially reduced Mg^+. Furthermore, we observe that higher order glymes as well as DMSO improve the solubility of Mg salts, but only the longer glyme chains reduce the dynamics of the ions in solution. This information provides critical design metrics for future electrolytes as it elucidates a close connection between bulk solvation and cathodic stability as well as the dynamics of the salt. Following this design matrix we compare the properties of different multivalent cations such as Mg^{2+}, Zn^{2+}, and Ca^{2+} which is essential to design new electrolytes for multivalent batteries.

References

[1] N. N. Rajput, X. Qu, N. Sa, A. K. Burrell, K.A. Persson, "The Coupling between Stability and Ion Pair Formation in Magnesium Electrolytes from First-Principles Quantum Mechanics and Classical Molecular Dynamics", Journal of the American Chemical Society, 2015.

[2] X. Qu, A. Jain, N. N. Rajput,Y.Zhang, S.P.Png, M. Brafman, L.Cheng, E. Maginn, L.A. Curtiss, K.A. Persson, "Electrolyte Genome Project: A Big Data Approach in Battery Materials Discovery", Computational Materials Science, 2015.

(496j) Bulk Silicon/Graphite Alloy Anode Material for High Energy Density Li-Ion Batteries

Lee, M., XALT Energy, LLC
Feng, P.,
Mahajan, V., XALT Energy, LLC
Dahlberg, K., University of Michigan
David, L.,
Albano, F., XALT Energy, LLC
Dhar, S., XALT Energy, LLC
Originally presented on: 11/11/2015 10:45:00 - 11:00:00

Silicon anodes are an optimum alternative to traditional graphite anodes because of their higher lithium ion storage capacity. However, the massive first cycle irreversible capacity loss and large fade during cycling have hindered its commercialization. In this paper, we present a solution for these issues in the form of commercial grade silicon/graphite alloy. A silicon-carbon alloy structure with carbon-coated silicon nanoparticles gives an ability to buffer volume changes and retain the silicon amorphous phase. The material in our approach achieved a first cycle reversible capacity of over 480 mAh/g when cycled between 0.01-1.5V at a c-rate of 0.05C, with first cycle efficiency of close to 95%. The capacity of the silicon alloy was ~30 mAh/g even at 10C rate after cycling, and recovered to a reversible capacity of 480 mAh/g at 0.05C. We performed further studies using galvanostatic intermittent titration (GITT) technique which showed the diffusion coefficient to be 10^{-9} to 10^{-12} cm^2/sec. Nyquist plots were obtained using electrochemical impedance spectroscopy (EIS) at every 5%SOC. 95x64 mm pouch cells (2Ah) with the silicon alloy coupled with a high capacity NMC cathode had impressive long term cycling, over 500 cycles, at a high c-rate (1C/1C), and stable c-rate performance at 10C. The cells were disassembled after cycling to study the SEI formation, electrode expansion, and coating stability. Physical measurement of electrodes and SEM analysis of active materials showed that even after long term cycling, the coating was intact without minor cracks or paste delamination.

Thermal Degradation of Aqueous Methyldiethanolamine (MDEA) with Continuous Injection of H_2S/CO_2 in High Pressure Reactor

*Priyabrata Pal and Fawzi Banat**
The Petroleum Institute, Abu Dhabi, United Arab Emirates
Email: fbanat@pi.ac.ae

Introduction

The removal of H_2S/CO_2 (known as acid gases) from a great variety of hydrocarbon rich natural gas is the main goal to increase the industrial and commercial utility of the hydrocarbon streams [1]. Generally, alkanolamine solvents are used to capture acid gases and there has been an increased attention for monitoring the volatile degraded products from the solvent and correlating with foaming issues to understand environmental emissions [2]. GASCO Company (Habshan, Abu Dhabi) is currently using aqueous methyldiethanolamine (MDEA, 40-50 wt%) to remove acid gases. Different researchers conducted thermal degradation studies to identify thermal degraded products from MDEA. One of the major operational problems facing the alkanolamine based absorption process is foaming. Foaming can be enhanced by introducing various contaminants like condensed liquid hydrocarbon, fine particles, surface active agents and alkanolamine degradation products [3]. Considering all these aspects, it is highly desirable to carry out thermal degradation of MDEA in presence of H_2S/CO_2 continuously and carry out foaming and corrosion studies with degraded MDEA to understand an accurate and complete knowledge of industrial process for the purification of natural gas that are technically, environmentally and economically more efficient.

Experimental

Experimental unit

Degradation studies were performed at 120°C and CO_2/H_2S partial pressure of 0.0675/2.025 bars having total pressure of 9.0 bars in presence of nitrogen for 865 hours. The gases are cooled down in the overhead condensers and the condensate is returned to the reactor. At the end of the degradation test, the samples was distilled in vacuum distillation unit and analyzed for high molecular weight degradation products.

Instrumentation

Ion chromatography (IC), proton-transfer-reaction quadrupole mass spectrometry (PTR-QMS), gas chromatography mass spectrometry (GC-MS), inductively coupled plasma optical emission spectrometry (ICP-OES) and titration equipment were used for analysis of the samples.

Results and Discussions

Analysis of initial MDEA solvent

The analyses of fresh and lean MDEA are presented in Table 1. Low levels of organic acids and other inorganic heat stable salts (HSS) in mg/kg were detected in fresh and lean MDEA.

Table 1 Analysis of fresh and lean MDEA

Parameter	Fresh MDEA	Lean MDEA
Sulfide as S, mg/kg	-	27
Mercaptan as S, mg/kg	-	< 0.01
CO_2, wt%	-	0.0008
Total heat stable salts, wt%	0.05	1.22
Water, wt%	49.4	55.5
pH	11.51	10.54
MDEA, wt%	49.9	40.1
Unidentified, wt%	0.7	2.3
Glycolic and lactic acid	45	540
Formic acid	100	860
Acetic acid	20	1250
Propionic acid	< 10	430
n-Butyric acid	< 10	160
Chloride	-	65
Phosphate	-	10
Sulfate	-	110
Thiosulfate	-	25
Suspended particles, ppmw	-	2

Kinetic studies of thermal degradation

The reaction rate was measured at 120°C and their reaction rate constants (k_A) were determined from first order kinetic equation as: $$\ln\left(\frac{C}{C_0}\right) = kt \quad (1)$$

where, k is first order rate constant. Table 2 summarizes the rate constant data obtained for MDEA degradation as well as formation of different degraded products observed in PTR-QMS with coefficient of regression values. Similarly, first order kinetics was also observed for MDEA degradation having rate constant to be 2.546 x 10^{-5} hr^{-1} [4]. It was observed that for formation of degraded compounds rate constant values are higher for fresh MDEA compared to lean solution.

Table 2 Rate constant of thermally degraded fresh and lean MDEA (in brackets)

Parameter	Methyl-diethanolamine	Methyl-ethanolamine	Dimethyl-acetamide	Dimethyl-aminoethanol	Diethanol-amine	Triethanol-amine
k x 10^4 (hr^{-1})	1.12 (1.82)	23.37 (12.05)	2.34 (1.62)	19.28 (5.9)	13.25	13.24 (4.02)
Coefficient of regression	0.9143 (0.9813)	0.9812 (0.9825)	0.9989 (0.9798)	0.9968 (0.9961)	0.9838 -	0.9799 (0.9398)

Organic acids in thermally degraded MDEA

Different organic acids such as formic, acetic, propionic, n-butyric, glycolic and lactic acids were present in lean MDEA with higher amounts while traces are present in fresh MDEA. Figure1 shows the formation of formic and acetic acid over degradation period of 865 hours. The rate of formation of formic acid on fresh MDEA is higher than acetic acid while for lean MDEA the formation rate was almost equal (Figure1).

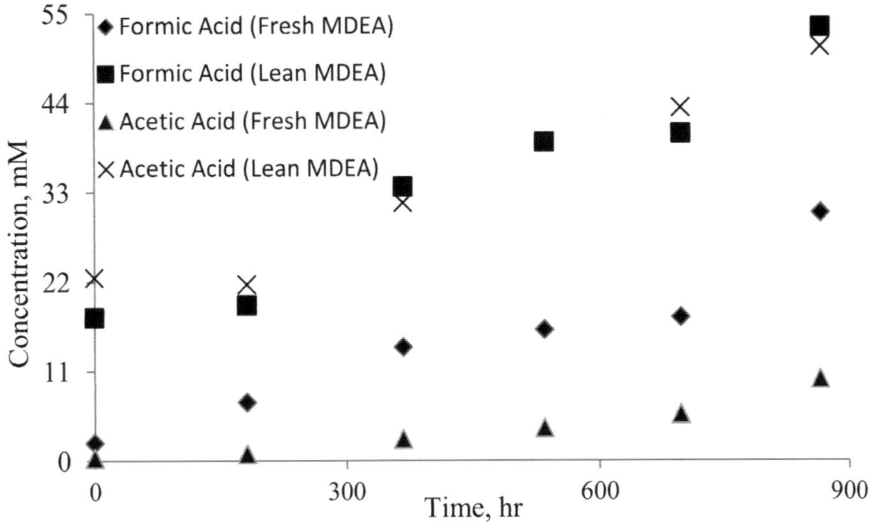

Figure1. Organic acids as HSS on thermal degradation of fresh and lean MDEA

Degradation products

It was observed that MDEA dissociated to diethanolamine (DEA) and finally monoethanolamine by delakylation/demethylation [5]. DEA was produced initially from MDEA fragmentation, in which the protonated MDEA was simply demethylated using the charge-remote fragmentation mechanism [6]. The second fragmentation involved the activation of the nitrogen atom with a hydrogen atom. This attack at the nitrogen atom causes one of the hydroxyethyl groups to leave the protonated MDEA to produce $[C_3H_9ON+H]^+$ (methylethanolamine). It was observed that after degradation of fresh and lean MDEA formation of DEA and methylethanolamine were relatively in same range with increasing degradation time as shown in Figure2. In this thermal degradation N,N-dimethylacetamide (DMA) was also identified. The molar concentrations of DMA increased with time and were greater for fresh MDEA as compared to lean MDEA. This may due to instability of the acetamide group which may further hydrolyze to produce organic acids.

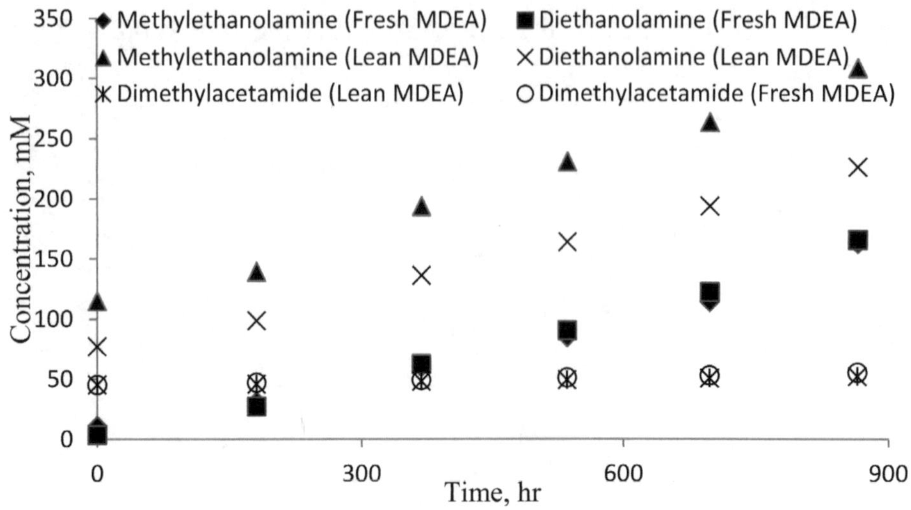

Figure2. Change of concentration with time for degraded products

Distillation and residue analyses

To determine the level of high-boiling degradation products, the solvent collected at the end of the degradation experiment was distilled using vacuum distillation. It was observed that residue content was almost twice for lean MDEA as compared to fresh MDEA (Table 3). The higher residue content indicated high molecular weight degradation products with higher concentration in lean MDEA.

Table 3 Distillation and residue analysis of fresh and lean MDEA

Distillation yields	Fresh MDEA	Lean MDEA
	wt%	wt%
Water fraction	45.80	53.3
MDEA fraction	47.90	41.0
Residue	1.44	2.74
Loss	4.82	3.03
Unidentified	> 90%	> 90%

Corrosion parameters

The corrosion potential for fresh and lean MDEA was measured using iron solubility test (IST; the ability of a solvent to dissolve iron in the solution) and complexing power (CP; ability of the solvent to keep metals in solution). The analysis of MDEA before and after degradation experiments are shown in Table 4. It was observed that lean MDEA has relatively higher IST value and can be well explained with higher acid content. However, the increase in corrosion potential with aging time is comparable with fresh MDEA. The low value of CP of the lean MDEA was not likely to have any corrosion problems in gas sweetening unit.

Table 4 Corrosion performance of fresh and lean MDEA

Parameter	Fresh MDEA	Lean MDEA

Aging time, hr	0	865	0	865
Iron solubility, ppm	5	119	225	292
Complexing power	low	low	low	low

Foaming tendency of solvents

The MDEA collected at the end of the degradation experiments were tested for foaming using N_2 [3]. The foam height and foam break time indicates relatively higher foaming tendency for degraded lean MDEA (Figure3). This could be due to higher heat stable salts present. The higher foaming tendency could also be related to potential hydrocarbons, emulsifiers etc. present in the solvent.

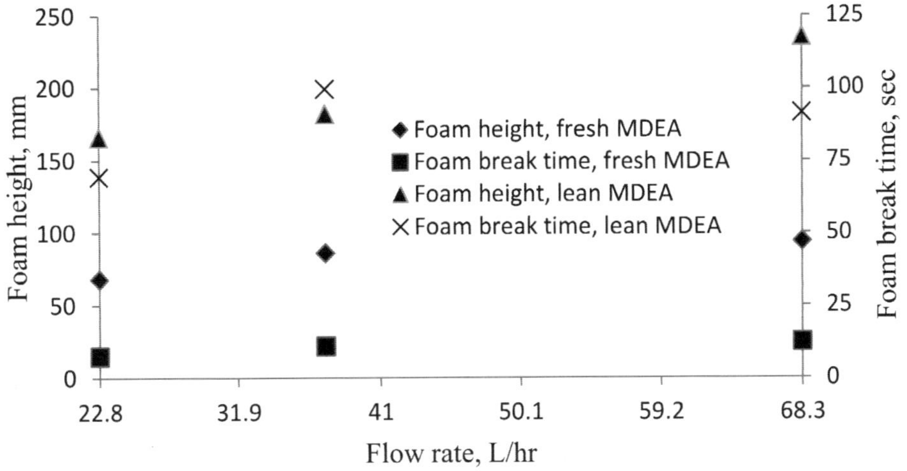

Figure3. Effect of nitrogen flow rate on foam height and foam break time

Conclusions

It was observed that thermal degradation of MDEA obeyed first order kinetics at 120°C. The major degradation products from both the fresh and lean MDEA were found to be similar. However, the concentrations of degradation products from lean MDEA were significantly higher. The higher acid content indicated oxidative degradation and possible oxygen ingress or presence of trace sulfur in the feed gas. The suspended particles of 2 ppmw indicated good solvent filtration efficiency. High IST value and lower CP of lean MDEA indicated lower corrosion potential of the solvent. However, the loss of MDEA due to degradation was significantly higher in the lean MDEA after degradation experiments. The lean MDEA sample showed relatively higher foaming tendency as seen from higher foam height and foam break time in the foaming tests.

Acknowledgement

The authors are grateful to The Petroleum Institute Gas Processing and Materials Science Research Center (Abu Dhabi) for funding the project (GRC 006), Shell Technology Centre Amsterdam (The Netherlands) and GASCO Company (Habshan, Abu Dhabi).

References

[1] Kohl, A.L. and Nielsen, R.B. *Gas purification*, 5th ed.; Gulf Publishing Company: Houston, TX, 1997.
[2] Mazari, S.A., Ali, B.S., Jan, B.M., Saeed, I.M., Nizamuddin, S. An overview of solvent management and emissions of amine-based CO_2 capture technology, *Int. J. Green. Gas Cont.*, 34(2015) 129-140.
[3] Alhseinat, E., Pal, P., Ganesan, A., Banat, F. Effect of MDEA degradation products on foaming behavior and physical properties of aqueous MDEA solutions, *Int. J. Green. Gas Cont.*, 37(2015) 280-286.
[4] Pal, P., AbuKashabeh, A., Al-Asheh, S., Banat F. Accumulation of heat stable salts and degraded products during thermal degradation of aqueous methyldiethanolamine (MDEA) using microwave digester and high pressure reactor, *J. Nat. Gas Sci. Eng.*, 21(2014) 1043-1047.
[5] Bedell, S.A., Worley, C., Darst, K. and Simmons, K., Thermal and oxidative disproportionation in amine degradation – O_2 stoichiomery and mechanistic implications, *Int. J. Green. Gas Cont.*, 5(2011) 401-404.
[6] Gross, M.L. Charge-remote fragmentation: an account of research on mechanisms and applications, Inter. *J. Mass Spectrom.*, 200(2000) 611–624.

419454 Process Design and Operational Strategy of CO2 Terminal from Recieving to Storage

Abstract:

As the carbon capture and storage (CCS) technologies are advanced, CO_2 transport options have been paid much attention. An infrastructure for transport including pipelines, ships, tanker truck, and railways will be collectively required to establish a large scale CO_2 transport network. Especially, ship transportation can play a central role in the early phase of CCS infrastructure development as it offers flexible routes between source and offshore storage sites. This work is focused on the design and operational strategy of the CO_2 terminal which acts as a connecting link between CO_2 liquefaction and the shipping section. Dynamic simulation is performed in order to have a more realistic understanding of the process. Four scenarios have been developed to define the operational strategy of the terminal: loading case, holding case, unloading case, and emergency shutdown. A reasonable equipment sizing has been done after consultation with vendor's data and available literature. The heat influxes to storage tank include the heat leak from the bottom surface, dry wall, and wet wall for the precise quantification of heat flow across the system. Reasonable control loops are appended to keep the process variables at their set points to allow the safe operation of the CO_2 terminal within the operational limits. Finally, a sensitivity analysis has been performed on some of the important design variables in order to understand their effect on the process performance and terminal operation. The results show that boil-off gas (BOG) generation within the CO_2 terminal depends on storage tank size, operating pressure, ambient conditions, insulation thickness, and the filling level of the vessel.

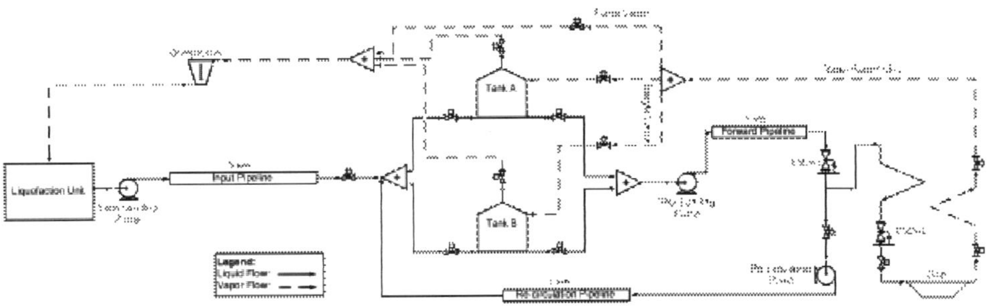

Basic Study on Development of Thermal Storage Technology at 500°C Using CaO /H_2O hydration reaction

Mr. KazukiKuwata, Nagoya University, Nagoya, JAPAN
Dr. Noriyuki Kobayashi, Nagoya University, Nagoya, JAPAN
Mr. Takehiro Esaki, Nagoya University, Nagoya, JAPAN
Mr. Daichi Iwase, Nagoya University, Nagoya, JAPAN
Dr. Hideki Kita, Nagoya University, Nagoya, JAPAN
Dr. Seiji Yamashita, Nagoya University, Nagoya, JAPAN

Introduction

Recently, global warming has become a big problem due to the massive consumption of fossil fuels. Effective use of exhaust heat produced in various industrial processes is a promise methodology to reduce the amount of energy consumption. For example, exhaust gas higher than 500 °C, especially from high temperature process such as firing furnace of ceramics, is emitted.

As a countermeasure, we focused on a chemical thermal storage which can store and reuse high temperature heat. A chemical thermal storage by using reversible thermal chemical reaction can solve the problem in imbalance between the heat supply and heat demand.

In this study, CaO/H_2O reversible chemical reaction was selected for its reaction system, since it has appropriate thermal operation range, high heat storage capacity and environmental affinity.In this reaction system, exothermic reaction occurs by hydration of CaO, and reverse endothermic reaction occurs by dehydration of $Ca(OH)_2$.

$$CaO(s) + H_2O(g) \leftrightarrow Ca(OH)_2(s) + 104.2\ kJ/mol$$

The durability performance for thousands of repetition, besides large thermal storage capacity and quick heat release and storage rate, is necessary to put chemical thermal storage into practical use. The durability performance of the CaO/H_2O reaction has been investigated in small scale of several grams or less of the reactant. For example, one hundred of stable reversible reaction is demonstrated in thermal gravimetric analysis study(1). Study on the influence of the cycle reaction in laboratory-scale reactor with hundreds grams of the reactant is, however, small in number and unclear.

In order to demonstrate the cycle reaction at the target temperature and investigate the effect of the cycle reaction on the reactant CaO, this study focused on heat release and storage cycle of the laboratory reactor (CaO 700 g in weight).The heat storage and release cycle was carried out 10 times repetition with packed bed type reactor. The effect of the cycle reaction was investigated by observing temperature response in the packed bed, visual evaluation of the reactant in the reactor, and the thermal gravimetric analyses.

Experimental setup

Equilibrium temperature of CaO /H_2O hydration reaction

Theoretical relation between temperature (T) of the reactor and water vapor pressure (P) of evaporator/condenser is able to describe by Clausius-Clapeyron equation. ΔH and ΔS are enthalpy and entropy in the course of the reaction, and R is gas constant.

$$\ln P = -\Delta H/RT + \Delta S/R$$

At the heat storage step, the heat at or higher than 370 °C is stored during dehydration reaction when the vapor pressure of the condenser is 3.2 kPa. At the heat release step, the heat at or higher than 500 °C is released in the hydration reaction when the vapor pressure of the evaporator is over 80 kPa. In addition, the transport rate of water vapor in the packed bed of CaO/$Ca(OH)_2$ will be accelerated when the vapor is under pressurized condition. Thus, this reaction system can store the heat at the target temperature of higher than 500 °C, and release the heat at the same temperature.

Experimental setup of cycle evaluation using packed bed type reactor

Figure 1 shows schematic of the experimental set up. A closed system of thermal storage apparatus was used for this experiment.This apparatus consists of a cylindrical reactor in electric furnace and an evaporator. The reactor was 106 mm in diameter, and 206mm in height. Two pipes were connected for vapor inlet and vapor outlet to the air. A vapor flow channel of 10 mm in outer diameter is located in the center of the cylindrical reactor so as to transport vapor from the center in the direction of the diameter during the hydration reaction.700 g of CaO particles, uniformed in 125 μm or less in diameter, were placed into the reactor and the packed bed was formed. The packed bed of CaO was 104mm in height. Four thermocouples T1, T2, T3, and T4 were placed inside the cylindrical packed bed. T1 and T2 were placed at 60 mm in the height, and at 30, 40 mm in the direction of the diameter from the axis. T3 and T4 were placed at 20 mm in the height, and at 30, 40 mm in the radial direction. Two pressure sensors were placed at the reactor where one was placed at the top of the reactor, and the other was placed at 50 mm in the direction of the radial direction.

Experimental procedure of cycle evaluation

At the heat release operation, the temperature of the packed bed type reactor was controlled at 230 °C. The pressure of the evaporator was set at 150 kPa. The hydration reaction began after the valve1 between the reactor and the evaporator was opened and vapor flowed into the reactor.

At the heat storage operation, the temperature of the packed bed type reactor was controlled at 550 °C. The dehydration reaction began under atmospheric condition after the valve2 was opened. The heat storage operation continued for 10~15 hours.

The heat storage and release cycle was carried out 10 times. After the 2 th, 5 th, and 10 th heat release operation, the conversion ratio of the reactant was quantified by measuring the reactor weight. And after 10 th heat release operation, visual evaluation of the reactant in the reactor was carried out.

Thermal gravimetric analysis

After the cycle evaluation, 10 mg of the reactant in the reactor was prepared as sample of thermal gravimetric analysis. The sample derived from the upper part of the reactant packed bed, and no pretreatment such as classification was carried out. Unreacted $Ca(OH)_2$ particles of 20~25 μm in diameter were also prepared for comparison. The repetitive reaction was carried out over the temperature condition ranging from minimum 220 °C for hydration to maximum 500 °C for dehydration. Partial pressure of the vapor was constant at 30 kPa, under N_2 gas atmosphere during the cycle of the reaction. The durability of the reaction was evaluated from the conversion ratio which was calculated from the change in the amount of the sample weight.

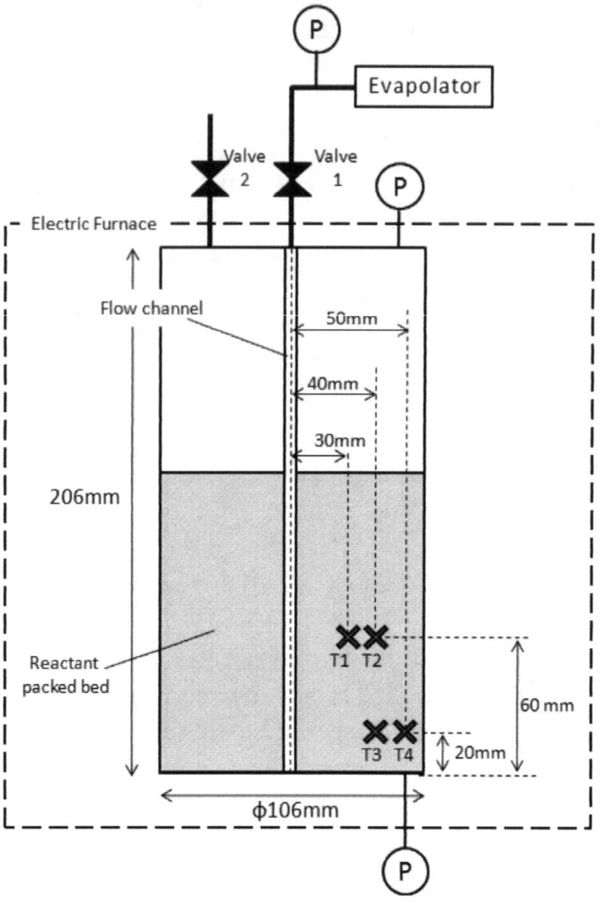

Figure 1. Schematic of the experimental apparatus

Results and Discussion

Cycle evaluation using packed bed type reactor

Figure 2 shows the temporal variation in the temperature of the packed bed and the pressure at heat release operation of the 10th cycle. After the reaction was started, the increase in the temperature was observed and the exothermic reaction was demonstrated. Each temperature measuring point reached maximum temperature within 2~3 minutes. The maximum temperature of T1 was 561 °C, and it was slightly higher than the equilibrium temperature 530 °C of 150 kPa. Then, the temperature of each measuring point decreased sequentially in the order of the distance to the reactor wall, attended with the thermal gradient in direction of diameter and axis. The pressure of the upper part and the bottom of the reactor leveled in a short time, and quick transport of the vapor in the reactant packed bed was confirmed. Hence, the temperature gradient in direction of radius and axis means that the heat transfer in the packed bed is slower than the heat release to the outside of the reactor.

Figure 3 shows the temporal variation in T2 at heat release operation 2 th, 5 th, 8 th, and 10 th cycle. The maximum temperature in the packed bed was higher than 500 °C at each cycle, so the heat release at the target temperature was demonstrated.

The change in the temperature behavior by reaction cycle was observed. The duration time of the maximum temperature tended to increase with increase of cycle number. After 8 th cycle, the change in temperature behavior was relatively small.

On the other hand, conversion ratio of the hydration reaction has little change although it rose slightly with increase of cycle number, as shown in Figure 4.

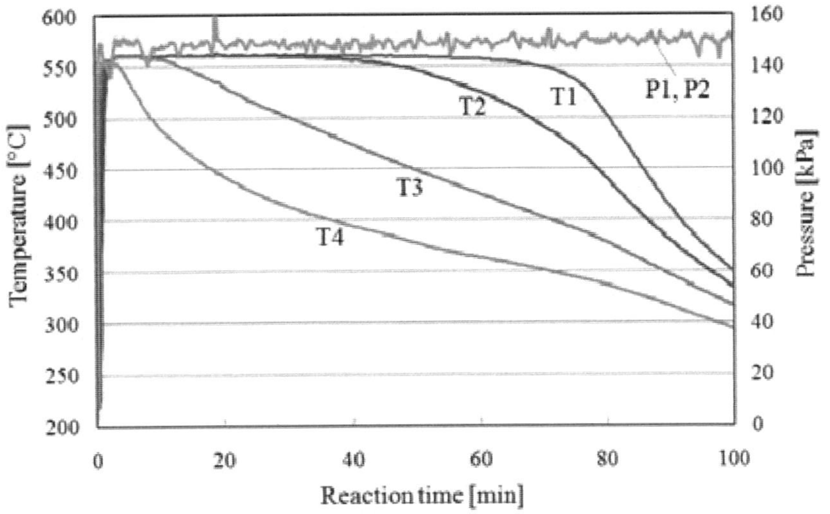

Figure 2. Temporal variation in the temperature of the packed bed and the pressure at heat release operation of 10 th cycle

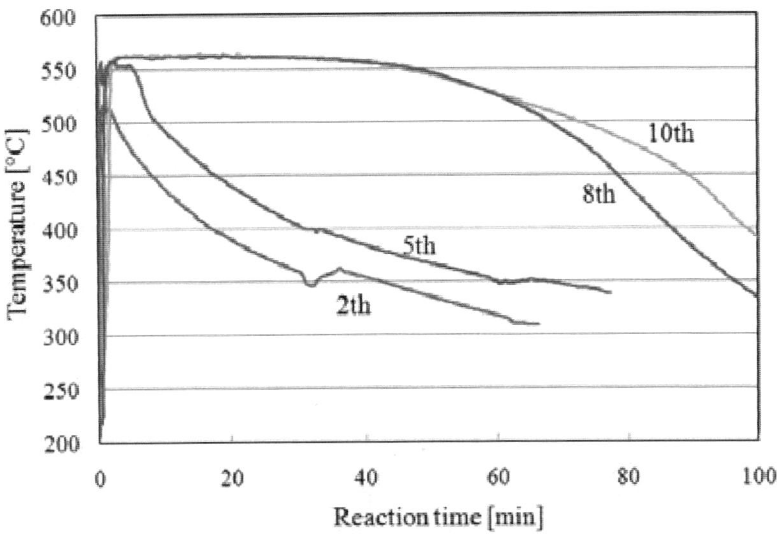

Figure 3. Temporal change in T2 at heat release operation of 2 th, 5 th, 8 th, and 10 th cycle

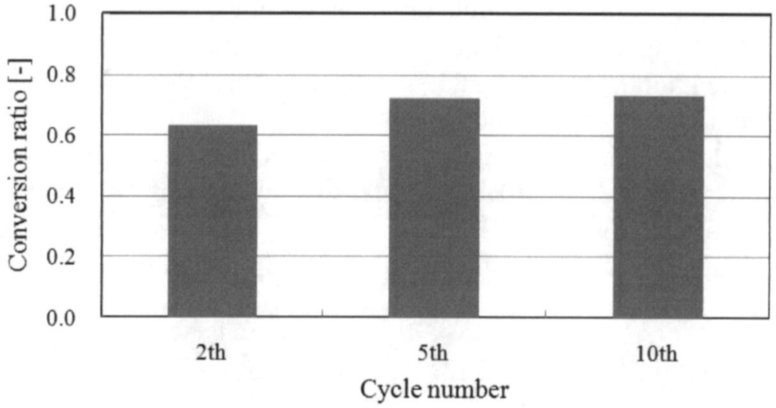

Figure 4. Conversion ratio for 2 th, 5 th, and 10 th cycle

Figure 5 shows the reactant in the reactor after 10 times of reaction cycle. The small reactant particle which was 125 μm or less in diameter grew up to several millimeter scale. These particles were not the perfect sintered particle, since these were fragile as it was possible to break by the hand easily. In addition, height of the packed changed from 104 mm to 146 mm. These structural changes of the reactant particle and the packed bed seemed to cause change of heat transfer rate in the reactant packed bed, and were reflected in the result shown in Figure 3.

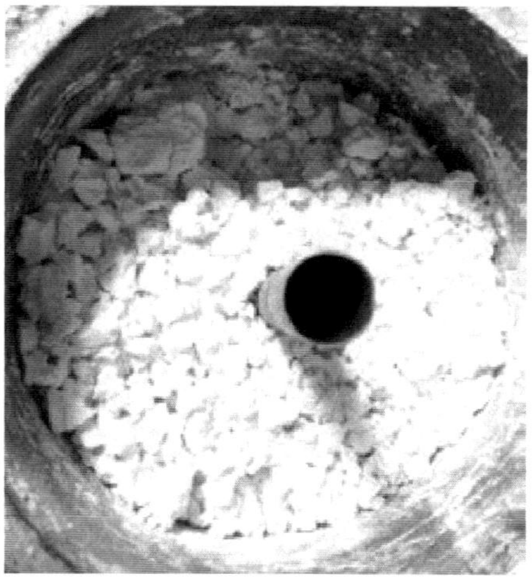

Figure 5. A snapshot of the reactant in the reactor after 10 times of reaction cycle

Reactivity of the reactant particle after the reaction cycle

Figure 6 shows the result of the thermal gravimetric analysis. The final reaction rate of the reactant derived from cycle evaluation was higher than 0.95 like the unreacted reactant. The time required to reach 0.8 was about 10 minutes. Therefore, no degradation due to sinter or side reaction was observed after 10 times of the heat storage and release cycle after long-time storage operation in atmospheric condition.

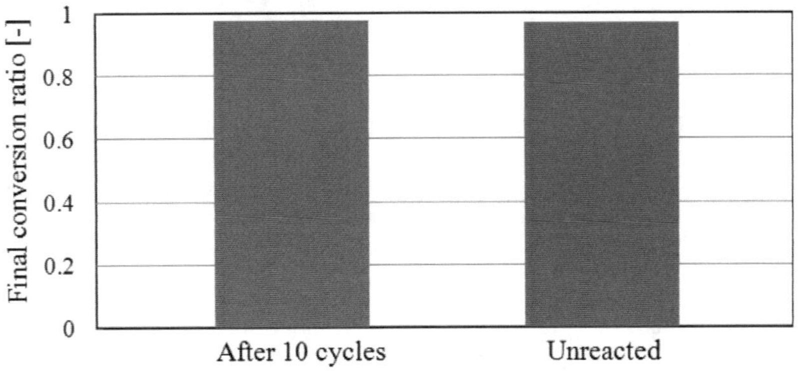

Figure 6. Result of the thermal gravimetric analysis

Conclusion

CaO/H_2O reversible chemical reaction was selected for chemical thermal storage which could release and store the heat over 500 °C.

In this study, heat release and storage cycle experiment by using the laboratory-scale reactor with CaO 700 g was carried to demonstrate the cycle reaction at the target temperature and investigate the effect of the cycle reaction on the reactant CaO.

Rise in the temperature up to 500 °C of the CaO packed bed inside the reactor was observed.

The behaviour of temperature variation in the packed bed over 10 time cycle was confirmed. This seemed to be caused by the structural changes of the reactant particle and the packed bed. From the results of the thermal gravimetric analysis, little degradation due to sinter or side reaction was confirmed after 10 times of the heat storage and release cycle.

It is suggest that this reaction system may be promised to realize a technology of heat storage and release on the level of 500 °C.

References

1. F. Schaube, L. Koch, A. Worner, and H. Muller-Steinhagen(2012), "A thermodynamic and kinetic study of the de- and rehydration of $Ca(OH)_2$ at high H_2O partial pressures for thermo-chemical heat storage," *ThermochemicaActa*, June, pp. 9-20.

424917 Design of Cascade Organic Rankine Cycle Utilizing LNG Regasification Process and CO2 Liquefaction Process

Abstract:

After the problems of global warming have been issued, development of environmental-friendly power generation attracted the attention of many researchers. One available technology for imtigating the global warming problem is CCS, which is composed of carbon capture, liquefaction of CO2, transportation and storage. This technology can reduce significant amount of CO2 emission but it has been reported that the CCS technology costs lots of energy so reducing the total energy assumption is neccessary for applying CCS to real plants.

Meanwhile, liquefied natural gas (LNG) is one of the most widely used form of fossil fuel with high levels of energy production and low greenhouse gas emissions. Also, due to the extremely low temperature of LNG, it can produce some additional electricty if there is a organic rankine cycle (ORC) utilizing the physical exergy of LNG. After the LNG imported is transported to regasificatioin termainal, it should be evaporated to be delivered to using site. Here, the regasification process can be utilized as heat sink of organic rankine cycle in order to produce some additional electiricy. Also, if the site of CO2 liquefaction process is placed near by the LNG regasification terminal, the CO2 can be adopted as heat source of the ORC, which uses the LNG cold exergy.

In this study, a cascade organic Rankine cycle which uses LNG cryogenic exergy was proposed. The proposed system utilizes LNG process as a heat sink so that it can produce additional power as well as obtain natural gas (NG), that is, LNG regasification process is adopted to produce additional power. For producing maximum amount of power, it is designed to be three stage cascade cycle. The first two stages are adopted to recover the LNG cold exergy directly and the third stage is used to recover additional exergy from the first two stages. Each stage uses binary working fluid to fit the temperature glide of LNG evaporation process. The best combination of working fluid is selected through irreversibility minimization by varying flow rate, composition and pressure of working fluids. After selecting the working fluids, process optimization is performed via a parametric study. The heat source is obtained from CO2 from the carbon capture process. Therefore the CO2 liquefaction process which is essential for CO2 capture process is substituted by the proposed process so it can reduce around 50% of CO2 liquefaction energy. An additional heat necessary to vaporize each working fluids is brought from free heat sources such as sea water. As a result, the proposed process produces 151.78 kJ/h kgLNG under a 25 heat sources and showed 21.85% second law efficiency without any additional energy injected except for LNG cold exergy.

421099 Development of Novel MoO2-Based Solid Oxide Electrolysis Cells (SOECs) for Syngas Generation from CO2 and H2O

Abstract:

Directly converting CO_2 and H_2O into syngas using intermittent renewable energy provides an alternative way to address our future oil depletion and climate change problems. Such novel technology can mitigate an unpredictable nature of renewable energy as well as its storage and transportation issues by converting carbon-free electrical energy from wind and solar into chemical energy stored in hydrocarbon fuels. In order to realize this CO_2-recycled fuel production process, one can utilize solid oxide electrolysis cells to convert CO_2 and H_2O into syngas using renewable electrical energy, which in turn can be converted to liquid fuels through a small modular Fisher-Tropsch (F-T) system.

Ni-based electrode is currently used for CO_2 and H_2O co-electrolysis process. However, the triple phase boundary, where the electrochemical reaction occurs for Ni-based electrode, is very limited which leads to a low electrochemical performance. In this study, MoO_2-based electrode was fabricated and applied to co-electrolysis. Unlike the Ni-based electrode, MoO_2-based electrode can promote the electrochemical reaction over its entire surface due to its ability to conduct both ions and electrons. Thus, it provides much higher electrochemically active surface area than that of Ni-based electrode, which should lead to its superior performance toward CO_2 and H_2O co-electrolysis. Corresponding voltage-current plots, gas outlet composition using GC/MS and electrochemical impedance spectroscopy (EIS) are analyzed to elucidate the reaction mechanism of CO_2 and H_2O co-electrolysis over MoO_2-based electrode.

436589 Mass Transfer Scale-up for Semi-Batch Polymer Production

Abstract:

For many polymerization processes it is of funamental importance to know the concentration of monomer at the polymerization site to determine, and tailor, the final properties of the polymer. This is particularly true is such processes in which three phases are involved, and the monomer needs to transfer from a gas phase to a liquid phase to the solid phase.

In this work, different approaches have been used to determine the concentration of the monomer in the diluent phase of the reaction, as a consequence of the existing mass transfer in the particular batch reactor of use. Namely, laboratory and pilot equipment of different sizes have been used to experimentally determine the gas-liquid mass transfer coefficient, and its dependency on different process parameters and geometry configurations. Based on literature relationships, the same mass transfer has been extrapolated to plant production conditions, and verified as well by meas of CFD simulations.

The results of the measurement and simulations at different scales have been put together to understand the governing factors in the reactor, and to be used for process and product development purposed.

It has been seen that the geometrical configuration of the different scale reactors if of fundamental importance to generate the right monomer concentration conditions, as well as to define and optimize the batch plant productivity.

423516 Design of a Large-Scale Micro-Channel Fischer-Tropsch Reactor: Model Development for Barrels-per-Day-Scale Reactor

Abstract:

In this study, a large-scale micro-channel Fischer-Tropsch (FT) reactor was designed. Based on 1-barrel per day of wax production rate, a model that solves momentum, material, and heat balance equations simultaneously was developed. Approximately 5,000 channels-1,000 for process and 4,000 for cooling channel-are decomposed into each unit cell, and every neighboring cells are then coupled together to solve the balance equations. The realistic flow distribution effect was incorporated into the model by using results obtained from computational fluid dynamics (CFD). The effect of several important variables, i.e., catalyst loading, coolant flow rate, channel layout, flow configurations, flow distribution, are clarified through the relevent case studies. The results of the model validation work, and some realistic issues that arises in the fabrication of the large scale reactor are presented as well.

435539 Effect of Friction Factor Correlations and Propagation Errors on Differential Pressure in a Crude OIL Fouling Measuring RIG

Abstract:

EFFECT OF FRICTION FACTOR CORRELATIONS AND PROPAGATION ERRORS ON DIFFERENTIAL PRESSURE IN A CRUDE OIL FOULING MEASURING RIG

Z. Tajudin+, *E.Diaz-Bejarano*+, *F. Coletti**, *S. Macchietto*+* *and G. F. Hewitt*+*

+*Department of Chemical Engineering, Imperial College London, South Kensington Campus, London SW7 2AZ, UK*

* *Hexxcell Ltd, Imperial College Incubator, Bessemer Building Level 2, Imperial College London, London SW7 2AZ, UK*

Corresponding author: s.macchietto@imperial.ac.uk

Keywords: *pilot scale, error of propagations, experiment, validation*

Abstract

In order to detect crude oil fouling experimentally, primary measurements of differential pressure and temperatures must be obtained with high fidelity, accuracy and reproducibility at (or close to) industrial conditions. Information of the thermal and hydraulic effects of fouling can be studied by using robust models to decouple the various phenomena involved. To start with, it is important to have a reliable set of primary measurements in which the robust model could be validated against the experiment data.

A novel pilot scale experimental facility for study of crude oil fouling, High Pressure Oil Rig (HIPOR), was developed and tested with non-fouling oil and crude oil at different operating conditions. The experiments have been carefully designed in order to develop comprehensive and reliable baseline validation for both oils without any fouling deposition. The model, an extension of Hexxcell's model for HiPOR, for the tubular test section with vertical geometries, considers the two dimensional (axial and radial) heat transfer in different domains, boundary conditions and other aspects (for example, heat losses) which are carefully determined.

The objectives of this work are divided into two. First, a predictive model for differential pressure of the HIPOR test section is validated against experimental results for both oils. The model consists three friction factors correlation was used to assist in the analysis and interpretation of the experimental data collected, compared with the variation of flowrate as well as different heat inputs. Second, the errors of propagation which are observed from the validation process have been identified and its sensitivity towards inlet bulk temperature increments quantified.

The validation process shows that both oils show excellent agreement with model predictions for at all ranges of operating conditions and capturing all responses including the transient processes. The best agreement is achieved for different friction factor correlations for each oil. The reported results also show that the deviation between simulation data and measured data due to the errors of propagation starts to increase when the inlet bulk temperature of oil is increased from 80 °C onwards. Furthermore, the errors of propagation have a linear correlation with the temperature increment.

It is concluded that the model of the HIPOR rig for non-fouling and crude oil conditions for differential pressure were successfully validated and error of propagations with temperature effects have been quantified.

Acknowledgments

This research was partially performed under the UNIHEAT project. The authors wish to acknowledge the Skolkovo Foundation and BP for financial support. ZT wishes to also acknowledge MARA and UniKL for their support and funding. The support of Hexxcell Ltd, through provision of Hexxcell Studio™, is also acknowledged.

References

S. Macchietto, G.F. Hewitt, F. Coletti, B.D. Crittenden, D.R. Dugwell, A. Galindo, G. Jackson, R. Kandiyoti, S.G. Kazarian, P.F. Luckham, O.K. Matar, M. Millan-Agorio, E.A. Müller, W. Paterson, S.J. Pugh, S.M. Richardson, D.I. Wilson, "Fouling in crude oil preheat trains: a systematic solution to an old problem", *Heat Transfer Engineering*, **32**, 3, 197-215 (2011).

F. Coletti and S. Macchietto, A dynamic, distributed model of shell-and-tube heat exchangers undergoing crude oil fouling. *Ind. Eng. Chem. Res.* **50** (8), pp 4515–4533 (2011).

Z. Tajudin, Experiments, Modelling and Validation of Crude Oil Fouling on Large Scale Rig. PhD, *Imperial College London*, UK – submitted for PhD awards (2015).

Hexxcell Ltd., 2015. Hexxcell Studio. http://www.hexxcell.com.

BITS Pilani
K K Birla Goa Campus
Goa, India
Department of Chemical Engineering

Experimental studies on unmixed combustion - packed bed reactor technology for heat transfer applications

AIChE Annual Meeting 2015
Salt Lake City, Utah, USA
(8th – 13th November 2015)

Session No.: 586a
Paper ID: 430702

Amol Deshpande
Srinivas Krishnaswamy
Krishnaswamy Ponnani

11th November 2015

Outline

innovate achieve lead

- Introduction / Motivation
- Objectives of the present work
- Experimental test rig
- Experimental methods and conditions
- Results and discussion
- Conclusion

BITS Pilani, K K Birla Goa Campus

Motivation

Issues with premixed combustion in power plants

1. Toxic emissions
2. Green House Gas (GHG) recovery from flue gas - major issue
3. Energy loss (through flue gas)
4. Inefficient heat transfer to water (Lower boiler efficiency) which results in larger equipment size and higher operating temperatures

Alternative to premixed combustion – **CLC (Chemical Looping Combustion)**

Unmixed combustion process

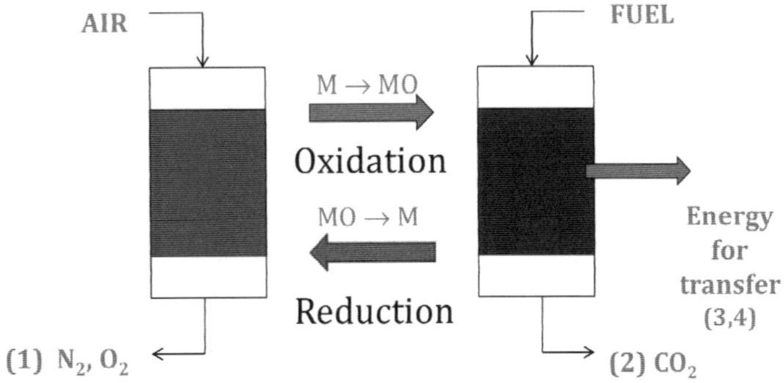

- OSRM (Oxygen Storage and Release Material)
  - Metal (M) / Metal Oxide (MO)
- Oxidation and Reduction of OSRM – Cyclic operation

References:

1. R. K. Lyon, N. J. Pittstown, U. S. Patent 005827496 (1998, Oct 27).
2. R. K. Lyon, J. A. Cole,, Combustion and Flame 121 (2000) 249-261.

Objectives

- To establish proof of concept test rig for UMC process based on packed bed reactor technology for heat transfer applications (Design, fabricate, install and commission)

- Study the effect of following parameters on heat transfer rate and bed temperature
  - Cycle time
  - Process air flow rate
  - Methane flow rate
  - Initial bed temperature

BITS Pilani, K K Birla Goa Campus

UMC for heat transfer applications

Experimental test rig

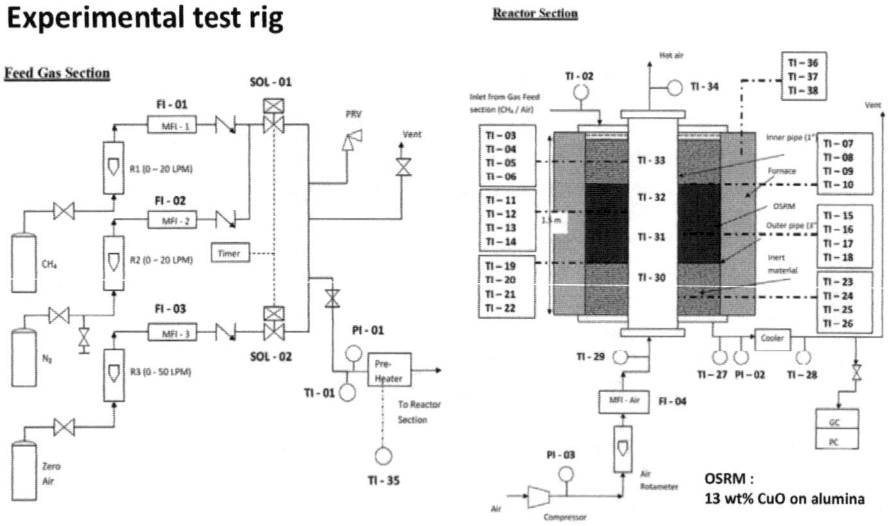

BITS Pilani, K K Birla Goa Campus

Experimental test rig (Photos)

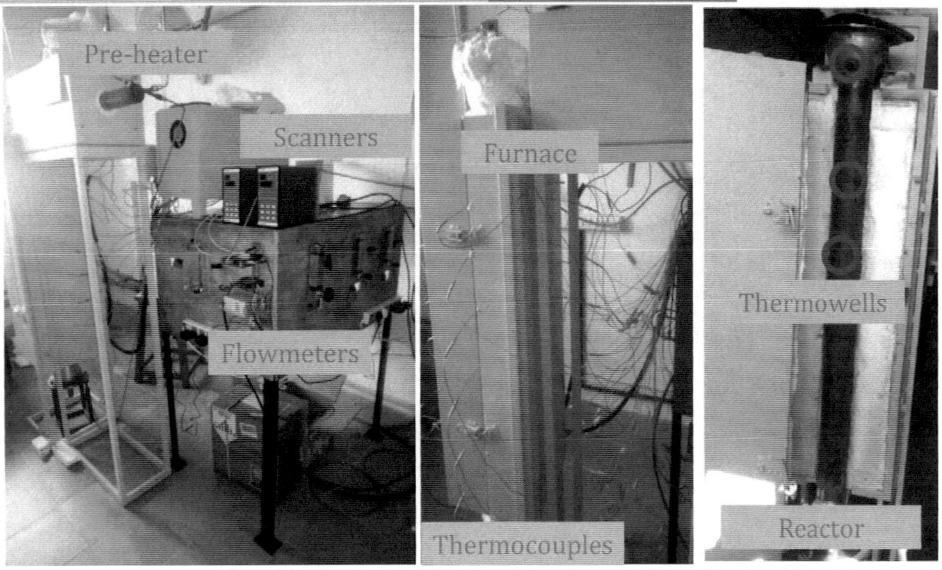

Cu/CuO as an OSRM

- Representative OSRM
- Ease of availability and relatively inexpensive
- Both oxidation and reduction reactions are exothermic

$4Cu + 2O_2 \rightarrow 4CuO$ ΔH_{rx}= - 312 KJ/mol

$4CuO + CH_4 \rightarrow 4Cu + CO_2 + H_2O$ ΔH_{rx}= - 178 KJ/mol

- Kinetics of oxidation and reduction of Cu based OSRM were determined using a new technique of pulsed differential micro-reactor (an alternative to TGA)

Engineering implications – Pulsed micro-reactor studies for estimating kinetics

- Cu/CuO was stable after being oxidized and reduced for over 50 cycles
- Material of construction of reactor needs to be chosen judiciously. SS reactor itself was observed to get oxidized at temperatures higher than 800° C
- During reduction using pure methane at temperatures above 700 ° C, presence of CO was observed in the product gases
- Coke formation was observed at higher temperatures when pure methane was used in reduction cycle

Experimental conditions

- **Amount of OSRM (13 wt% CuO on alumina) used = 500 gm**

Parameters	Oxidation	Reduction
Initial bed temperature (K)	873	873
Inlet temperature (K)	873	873
Reactant flow rate (LPM)	30 (Air)	4 (CH_4) 16 (N_2)
Cycle time (s)	300	180
Coolant air inlet temp (K)	303	303
Coolant air inlet pressure (bar(g))	1.5	1.5
Inner air flow rate (LPM)	75	75

Results – Bed temperature profile (For 2 cycles)

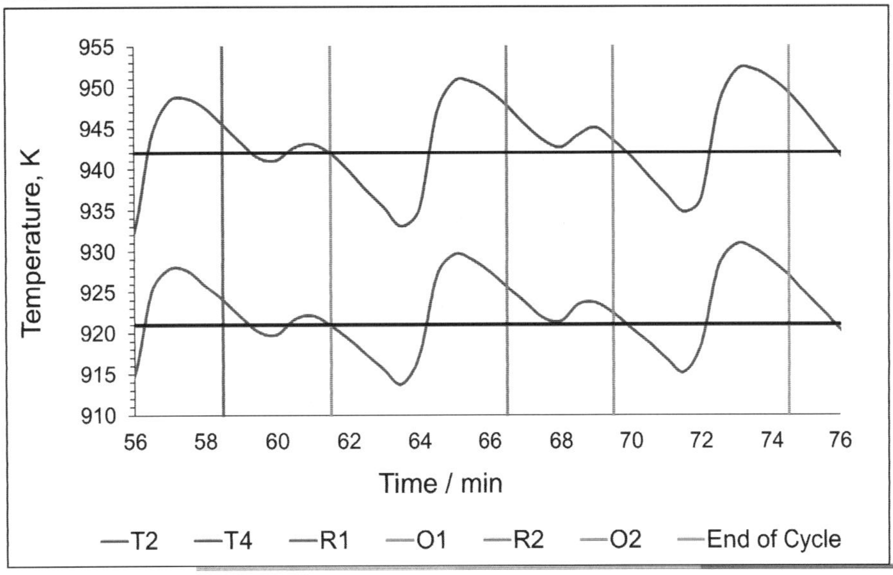

BITS Pilani, K K Birla Goa Campus

Results – Amount of radial heat transfer

Sr No	Process air flow rate (LPM)	Methane flow rate (LPM)	N2 flow rate (LPM)	Initial bed temperature (K)	% of Radial heat transfer
1	10	4	16	873	58 ± 1
2	20	4	16	873	61 ± 1
3	30	4	16	873	64 ± 1
4	40	4	16	873	63 ± 1
5	30	2	18	873	47 ± 1
6	30	4	16	873	64 ± 1
7	30	6	14	873	58 ± 1
8	30	4	16	823	55 ± 1
9	30	4	16	873	64 ± 1
10	30	4	16	898	57 ± 2

BITS Pilani, K K Birla Goa Campus

Effect of cycle time

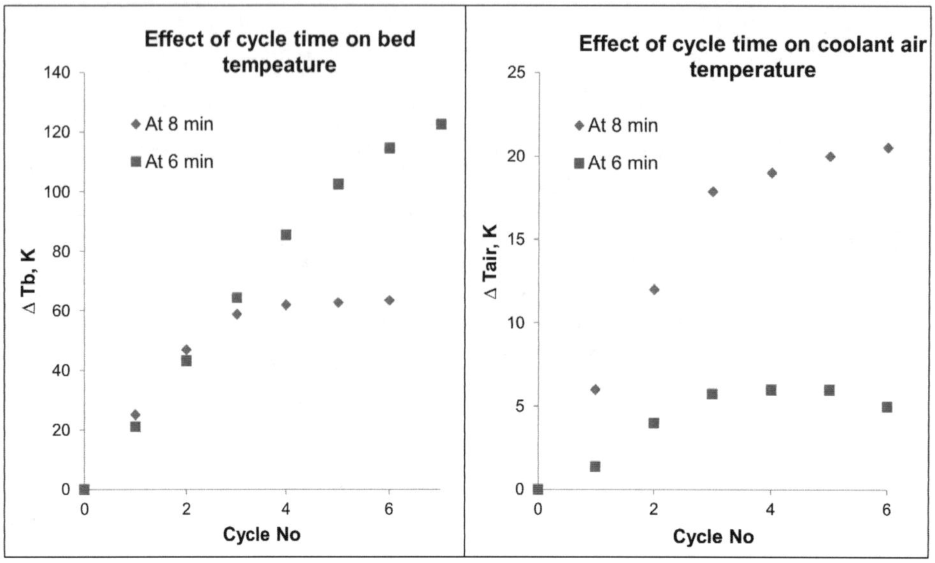

Effect of process air flow rate

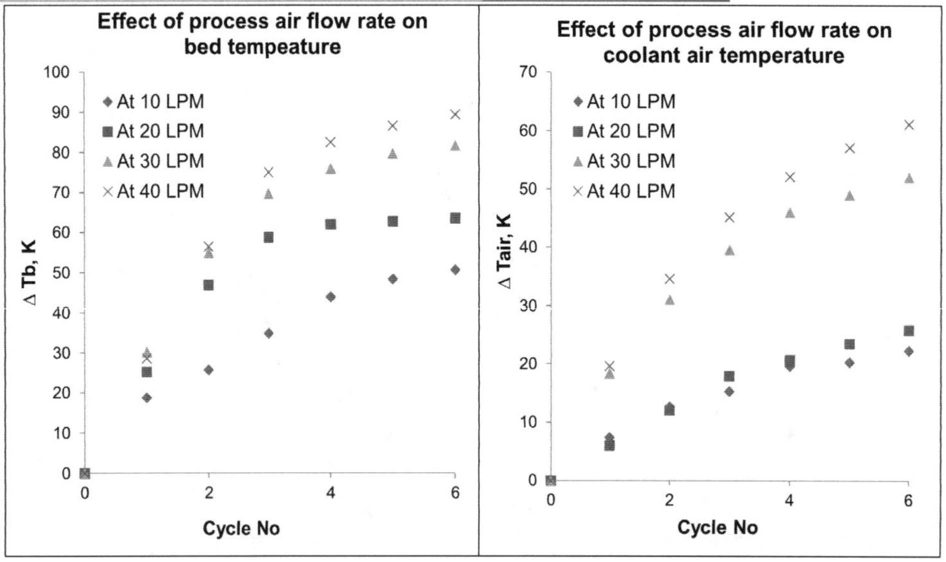

Effect of initial bed temperature

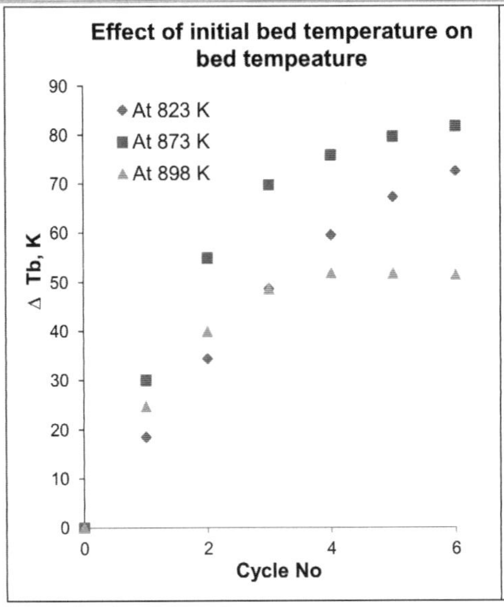

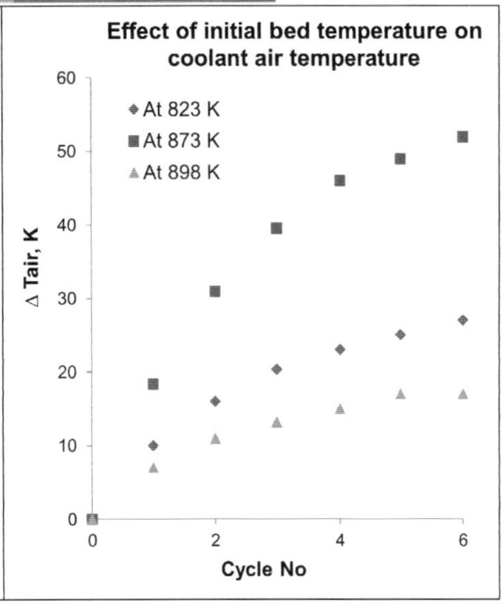

BITS Pilani, K K Birla Goa Campus

Effect of methane flow rate

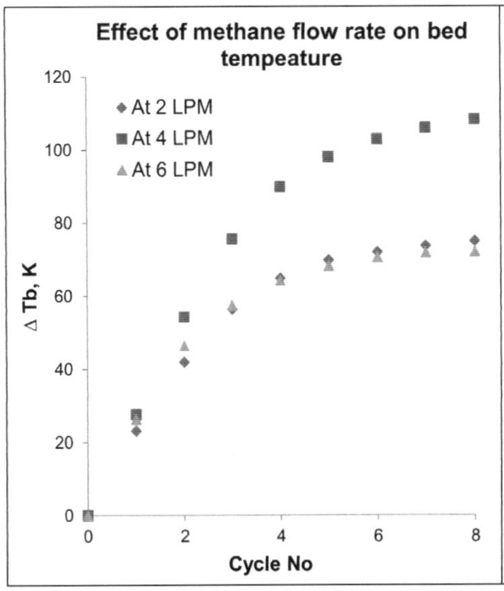

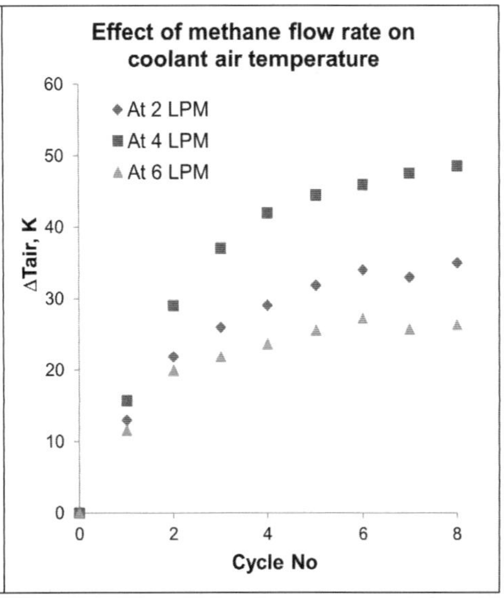

BITS Pilani, K K Birla Goa Campus

Conclusion

- Unmixed combustion process based on packed bed reactor technology can be effectively used for heat transfer applications
- Cu/CuO was found to be stable and favorable material for such applications.
- Near isothermal conditions in the bed can be maintained by adjusting cycle time
- Amount of radial heat transfer can be increased by selecting appropriate reactant flow rates and initial bed temperature

References

1. R. K. Lyon, N. J. Pittstown, Methods and systems for heat transfer by unmixed combustion, U. S. Patent 005827496 (1998, Oct 27).
2. R. K. Lyon, J. A. Cole, Unmixed combustion: An alternative to fire, Combustion and Flame 121 (2000) 249-261.
3. S. Noorman, M. van Sint Annaland, H. Kuipers, Packed bed reactor technology for chemical-looping combustion, Ind. Eng. Chem. Res. 46 (2007) 4212-4220.
4. S. Noorman, M. van Sint Annaland J. A. M. Kuipers, Experimental validation of packed bed chemical-looping combustion, Chemical Engineering Science. 65 (2010) 92-97.
5. F. Garcı´a-Labiano, L. F. De Diego, J. Ada´nez, A. Abad, P. Gaya´n, Reduction and oxidation kinetics of a copper-based oxygen carrier prepared by impregnation for chemical-looping combustion, Ind. Eng. Chem. Res. 43 (2004) 8168-8177.
6. S. Noorman, F. Gallucci, M. van Sint Annaland, J. A. M. Kuipers, Experimental Investigation of Chemical Looping Combustion in Packed Beds: A Parametric Study, Ind. Eng. Chem. Res. 50 (2011) 1968-1980.

innovate achieve lead

Acknowledgements

- Department of Science and Technology (DST), Govt of India
- Thermax Ltd., Pune, India
- BITS Pilani K.K. Birla Goa Campus, Goa

innovate achieve lead

Thank you

Questions??

BITS Pilani, K K Birla Goa Campus

424470 Numerical Investigation and Optimization of Flow and Heat Transfer in a Multi-Tubular Fixed Bed Reactor

Abstract:

Numerical investigation and optimization of flow and heat transfer in a multi-tubular fixed bed reactor

Li Hao, Bin Jiang, Luhong Zhang

Abstract: Multi-tubular fixed-bed reactor is one of the key equipment to propylene oxidation process and the reactor performance is closely related to heat carrier flow, heat transfer and temperature distribution. Reasonable reactor structure and suitable flow and temperature field distribution can effectively control hot spot temperature and improve the efficiency of the reactor. In this paper, a numerical study of fluid flow and heat transfer performance in the shell side of an industrial-scale cross-flow multi-tubular fixed bed reactor for propylene oxidation reaction was presented using computational fluid dynamics (CFD) method. Detailed information of fluid field and heat transfer in the shell-side of the industrial scale multi-tubular fixed-bed reactor was obtained by using three steps CFD simulation analysis method to control hot spot temperature. Firstly, to simulate flow behaviors, catalytic oxidation reaction, heat and mass transfer adopting porous medium model on tube side to achieve the temperature distribution. Secondly, based on the conclusions of tube-side, a novel configuration multi-tubular fixed-bed reactor was proposed and their performance of fluid flow and heat transfer was analyzed to ensure the uniformity condition on shell-side. There were circulation channels in the shell-side and there was no tube in the central region. Finally, the effects of operating conditions and structural parameters to control the hot spots were investigated, including inlet molten salt flow rate, flow direction, baffles number, baffles cut area, non-tube central region and flow circulation channels. And the optimization parameters were obtained. The simulation results showed that the velocity and temperature distribution varied along axial and radial direction, and there were stagnation regions before and behind the baffles, in which the heat transfer coefficient was low, but high heat transfer was observed in cross flow zones and baffle cut zones. The central region spot position decreased significantly with the central region area increased. The radial temperature difference increased with the increase in the number of circulation channels. The validity of the CFD results was confirmed by comparing its results of outlet temperature, total pressure drop and average heat transfer coefficient with empirical method results under different heat flux and inlet velocity conditions. The calculating results had a good agreement with the results of empirical methods. And this study had significant theoretical and application meanings for the design of multi-tubular fixed-bed reactors.

Keywords: Numerical analysis; multi-tubular fixed-bed reactor; structural optimization; heat transfer; hot spot control

(586c) First Principles Based Design of 3D Pyrolysis Reactors Using LES

Van Cauwenberge, D. J., Universiteit Gent
Reyniers, P. A., Ghent University
Floré, J.,
Van Geem, K. M., Ghent University
Marin, G. B., Ghent University
Originally presented on: 11/11/2015 15:49:00 - 16:06:00

Pyrolysis is the main process for the production of many valuable organic building block chemicals such as ethylene, propylene and vinyl chloride. The free-radical gas phase reactions are accompanied by secondary reactions leading to the formation of a carbonaceous coke layer on the inner walls of the reactor tubes. This layer leads to an increased pressure drop over the reactor causing a loss in selectivity. Additionally, the insulating effect of the layer forces a higher furnace firing rate and higher tube metal temperatures to maintain the same conversion. In industrial plants regular decoking procedures are hence inevitable.

Because of this detrimental effect on the overall economics, the design of novel reactor geometries exhibiting improved heat transfer and reduced metal temperatures with the aim of increasing run lengths has received quite some attention [1]. In order to assess the effect of these geometries however, one must account for the chemical kinetics, as well as the complex flow phenomena in the reactor. While reliable microkinetic models with thousands of reactions and hundreds of intermediates are widely applied, implementing these efficiently in a Computational Fluid Dynamics (CFD) code remains a challenge [2]. Commercial packages such as ANSYS Fluent for example do not even allow kinetic models with more than 50 species.

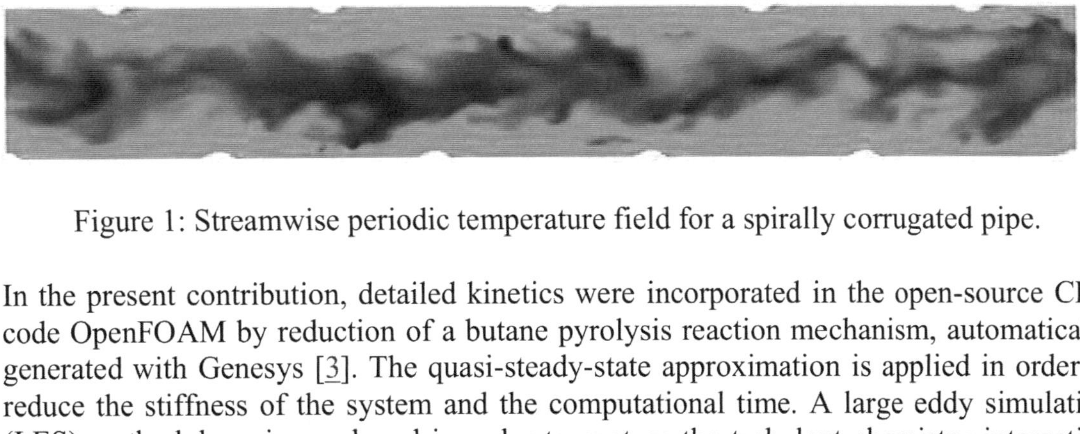

Figure 1: Streamwise periodic temperature field for a spirally corrugated pipe.

In the present contribution, detailed kinetics were incorporated in the open-source CFD code OpenFOAM by reduction of a butane pyrolysis reaction mechanism, automatically generated with Genesys [3]. The quasi-steady-state approximation is applied in order to reduce the stiffness of the system and the computational time. A large eddy simulation (LES) methodology is employed in order to capture the turbulent-chemistry interaction on a smaller scale than the classical Reynolds-averaged Navier-Stokes (RANS) turbulence models. The length of the computational domain is limited by extending on the periodic methodology of Patankar et al. [4] for fully developed velocity and temperature fields. This novel approach is validated by comparison with full-scale simulations for a variety of geometries and Reynolds numbers. Additionally, a pilot plant setup with a standard tubular reactor and a spirally corrugated tube is simulated and compared with the experimental results. The advantages and disadvantages of using three-dimensional reactor configurations instead of conventional tubular reactors are discussed.

Citations

[1] C.M. Schietekat, D.J. Van Cauwenberge, K.M. Van Geem, G.B. Marin, Computational fluid dynamics-based design of finned steam cracking reactors, AIChE J., 60 (2013).

[2] K. He, M.G. Ierapetritou, I.P. Androulakis, Integration of on-the-fly kinetic reduction with multidimensional CFD, AIChE J., 56 (2010) 1305-1314.

[3] N.M. Vandewiele, K.M. Van Geem, M.-F. Reyniers, G.B. Marin, Genesys: Kinetic model construction using chemo-informatics, Chem. Eng. J., 207?208 (2012) 526-538.

[4] S.V. Patankar, C.H. Liu, E.M. Sparrow, Fully Developed Flow and Heat Transfer in Ducts Having Streamwise-Periodic Variations of Cross-Sectional Area, J. Heat Transfer, 99 (1977) 180-186.

Acknowledgements

DVC gratefully acknowledges financial support from the combined IWT and BASF Antwerpen N.V. Baekeland program. PAR acknowledges financial support from a doctoral fellowship from the Fund for Scientific Research Flanders (FWO). The authors also acknowledge the financial support from the Long Term Structural Methusalem Funding by the Flemish Government ? grant number BOF09/01M00409. The computational work was carried out using the STEVIN Supercomputer Infrastructure at Ghent University, funded by Ghent University, the Flemish Supercomputer Center (VSC), the Hercules Foundation and the Flemish Government ? department EWI.

East China University of Science and Technology

Numerical Study on Enhancement of Cracking Reaction Employing a Novel Coil

Dehong Bai, Yuan Zong, Ling Zhao

State Key laboratory of Chemical Engineering

East China University of Science and Technology

Shanghai, China, 200237

November the 11th, 2015

15AIChE

Annual meeting, Salt lake city, UT

Outline

1. *Background*
2. *Mathematical model*
3. *Results and discussion*
4. *Overall evaluation*
5. *Conclusions*

Background

➢Process analysis

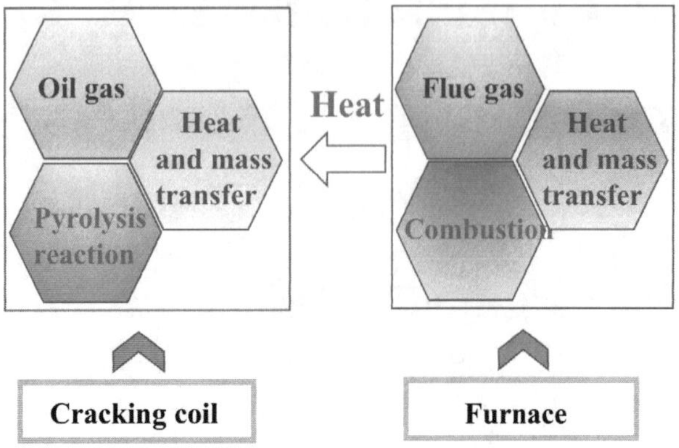

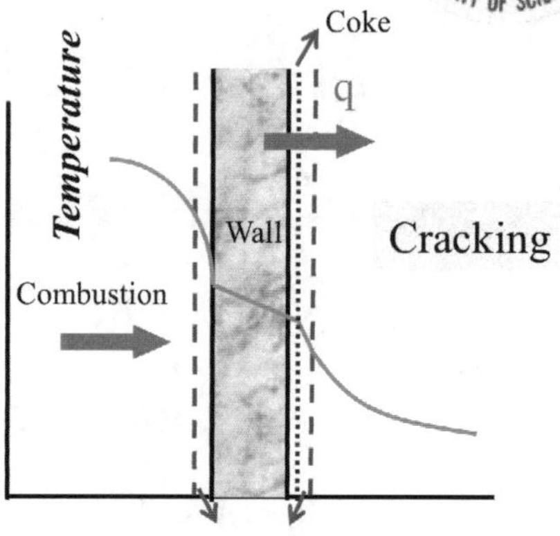

Boundary layer

☐ **Highly coupled process**: flowing, mass and heat transfer (Physical) ⇄ Interaction ⇄ reaction (Chemical).

✓ Research goal: **reducing boundary layer and reinforce central mixing to improve olefin yield as well as extending operating cycle.**

1

Background

➢Transfer enhancement

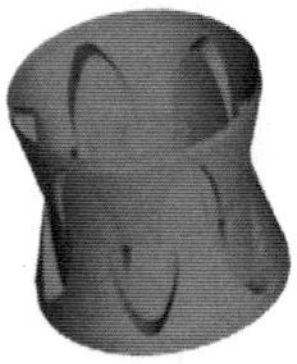

Hollow cross disk (HCD)

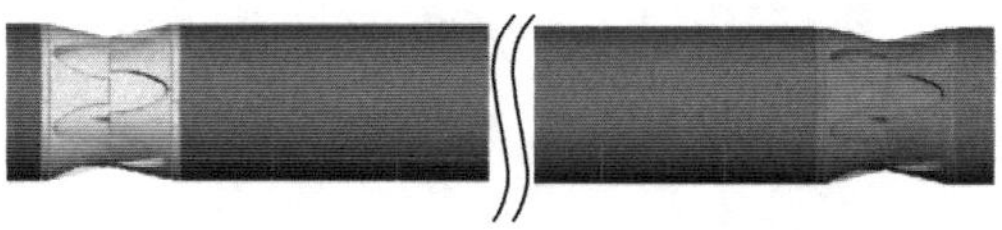

Improved coils setup

☐ **Advantages:**

- transfer enhancement
- comparatively low cost
- convenient for installation

☐ **Disadvantages**:

- additional pressure cost
- limited acting range
- difficult to manufacture

✓ ***Structure optimization: redistribute temperature and concentration profiles with suitable pressure cost***

2

Mathematical model

➢Governing equations:

Initializing

□ **Coupled algorithm**

Transfer conservation: $div(\rho u\phi) = div(\Gamma grad\phi)$

Turbulent :

$$\boldsymbol{k}: \frac{\partial}{\partial x_j}(\rho u_j k) = \frac{\partial}{\partial x_j}\left[\left(\mu + \frac{\mu_t}{\sigma_k}\right)\frac{\partial k}{\partial x_j}\right] + G_k - \rho\varepsilon$$

$$\boldsymbol{\varepsilon}: \frac{\partial}{\partial x_j}(u_j \varepsilon\rho) = \frac{\partial}{\partial x_j}\left[\left(\mu + \frac{\mu_t}{\sigma_\varepsilon}\right)\frac{\partial \varepsilon}{\partial x_j}\right] + \frac{\varepsilon}{k}(C_1 C_k - C_2\varepsilon\rho)$$

Species transport:

$$\frac{\partial \rho u_j Y_i}{\partial x_j} = \frac{\partial}{\partial x_j}\left[\left(\rho D_{i,m} + \frac{\mu_t}{Sc_t}\right)\frac{\partial y_i}{\partial x_j}\right] + R_i$$

$$EDC: R_i = \frac{\rho(\varepsilon^*)^2}{\tau^*[1-(\tau^*)^3]}(Y_i^* - Y_i)$$

Convergence

N

Y Output

3

Mathematical model—*Numerical* details

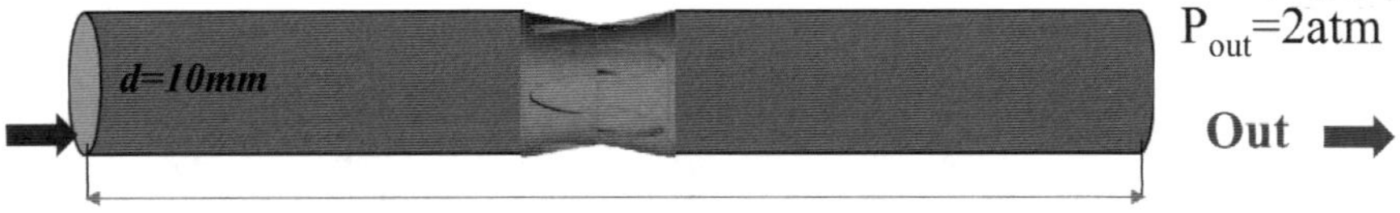

Cracking kinetic[2]	A (s^{-1} or $m^3/(mol \cdot s)$)	E/(J/mol)
$C_3H_8 \longrightarrow C_2H_4 + CH_4$	4.692×10^{10}	2.115×10^5
$C_3H_8 \leftrightarrow C_3H_6 + H_2$	5.888×10^{10}	2.144×10^5
$C_3H_8 + C_2H_4 \longrightarrow C_2H_6 + C_3H_6$	2.536×10^{13}	2.469×10^5
$2C_3H_6 \longrightarrow 3C_2H_4$	1.514×10^{11}	2.332×10^5
$2C_3H_6 \longrightarrow 0.5C_6 + 3CH_4$	1.423×10^{9}	1.902×10^5
$C_3H_6 \longrightarrow C_2H_2 + CH_4$	3.794×10^{11}	2.482×10^5
$C_3H_6 + C_2H_6 \longrightarrow C_4H_8 + CH_4$	5.553×10^{14}	2.508×10^5
$C_2H_6 \leftrightarrow C_2H_4 + H_2$	4.652×10^{13}	2.725×10^5
$C_2H_4 + C_2H_2 \longrightarrow C_4H_6$	1.026×10^{12}	1.725×10^5

[1] P.S. Van Damme, S.Narayanan, G.F. Froment. AIChE journal, 1975, 21(6): 1065-1073.

[2] K.M. Sundaram, G.F. Froment. Chemical Engineering Science, 1977, 32(6), 601-608.

4

Mathematical model

Table I: discretization schemes for different items

Pressure-velocity coupling	Coupled
Pressure	Second order
Density	Second order upwind
Momentum	Second order upwind
Turbulent kinetic energy	Second order upwind
Turbulent dissipation rate	Second order upwind
Species	Second order upwind
Energy	Second order upwind

➢comparing calculation:

Case I: smooth pipe 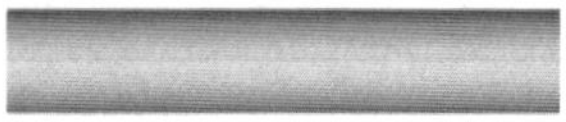 **(Smooth)**

Case II: novel coil  **(1-HCD)**

5

Results and discussion

➢Flow pattern:

Case I **Case II**

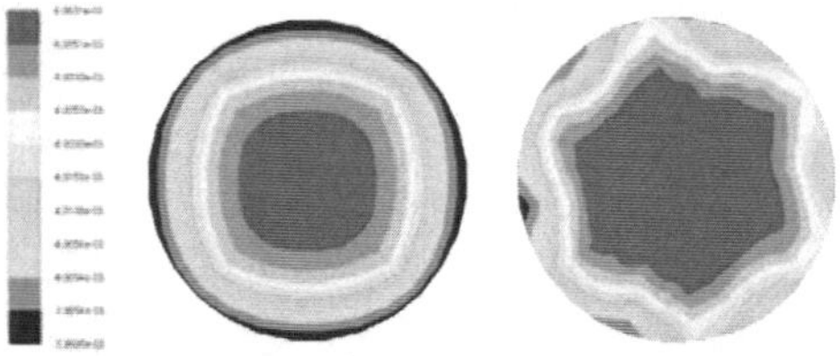

mole concentration of C_2H_4

Case I **Case II**

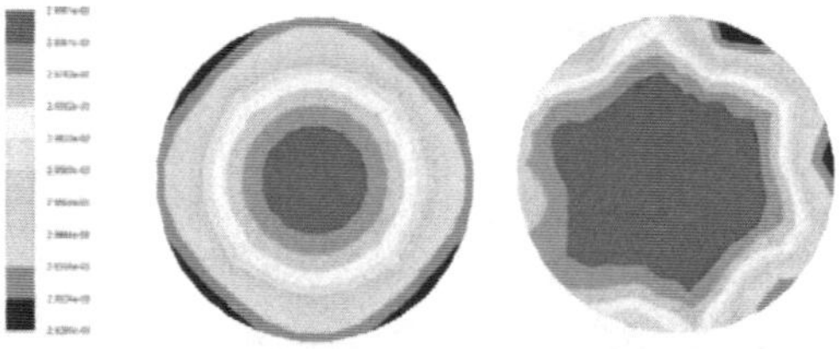

mole concentration of C_3H_6

- □ **Obvious concentration improvement of C_2H_4 and C_2H_6**

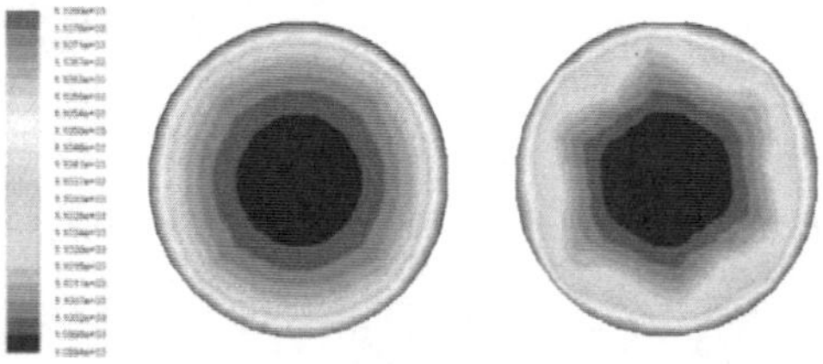

Temperature

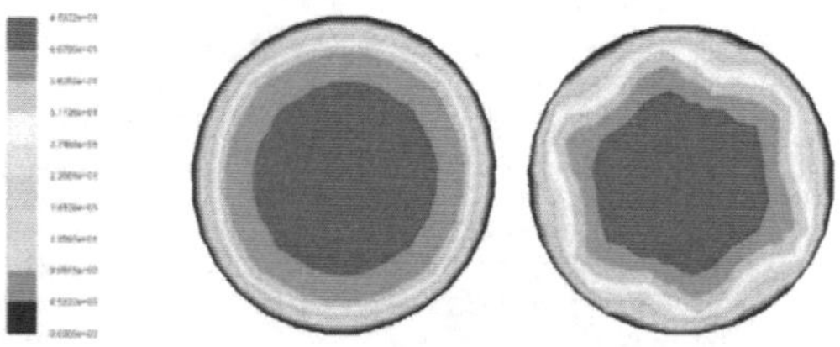

velocity

- □ **Apparent increase of velocity in the wall vicinity.**

6

Results and discussion

➢Product distribution:

Table II: Complete product profile smooth pipe and novel coil

components	reference	smooth	1-HCD
C_3H_8	7.4286	7.9634	7.2410
CH_4	17.7857	16.1792	14.5023
C_3H_6	10.5	14.5012	21.1852
H_2O	28.5715	28.5688	28.5349
C_6	4.5	2.5605	1.5838
H_2	1.2857	1.2146	1.3262
C_2H_6	2.5	2.4769	1.0388
C_2H_4	25	25.4040	23.9670
C_2H_2	0.3571	1.1284	0.6210
C_4H_8	0.4286	5.5612e^{-5}	1.0236e^{-5}
C_4H_6	1.6428	2.7700e^{-3}	5.3234e^{-4}

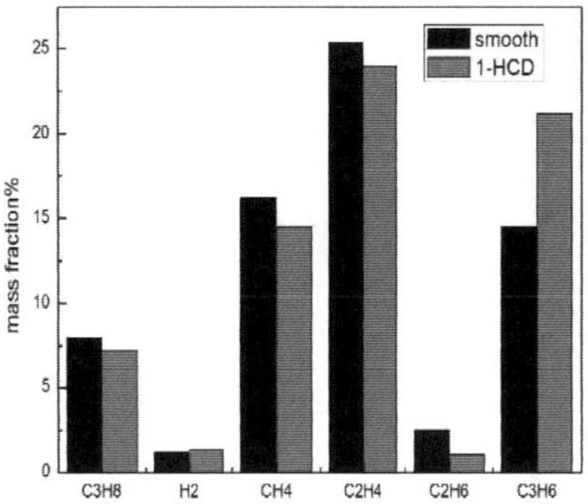

Comparison of mass fraction of major products in different cases

Table III: General reaction results

	T/k	Δp/Pa	$\beta_{(C2H4)}$%	$\beta_{(C3H6)}$%	$\beta_{(olefin)}$%	$X_{(C3H8)}$%
smooth	1090.54	22594.14	40.255	22.978	63.233	88.795
HCD	1092.77	24031.06	34.548	33.190	67.738	89.812

7

Results and discussion

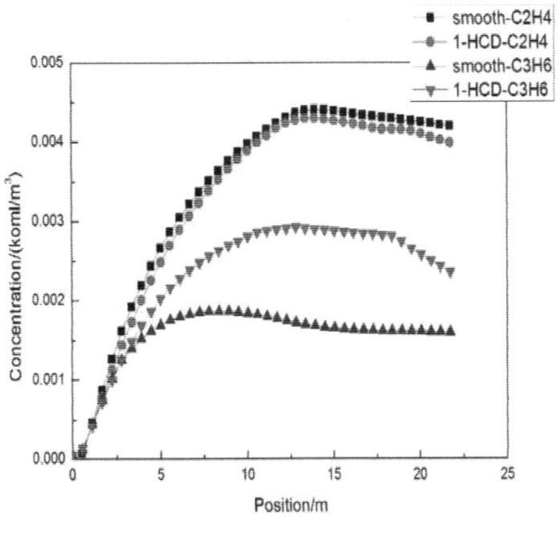

Concentration of C_2H_4 and C_3H_6

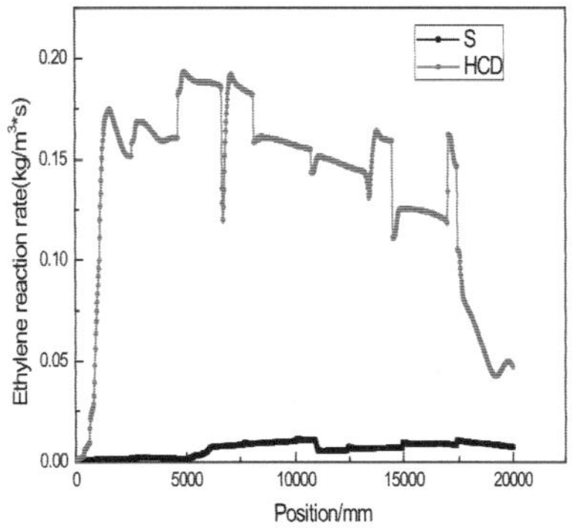

Reaction rate of C_2H_4

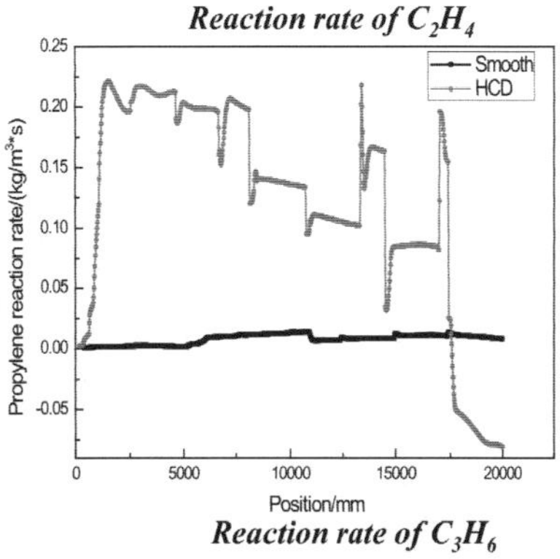

Reaction rate of C_3H_6

✓ **Dramatic contrasts demonstrate the effectiveness of HCD internal.**

8

Results and discussion

➢Coking suppression :

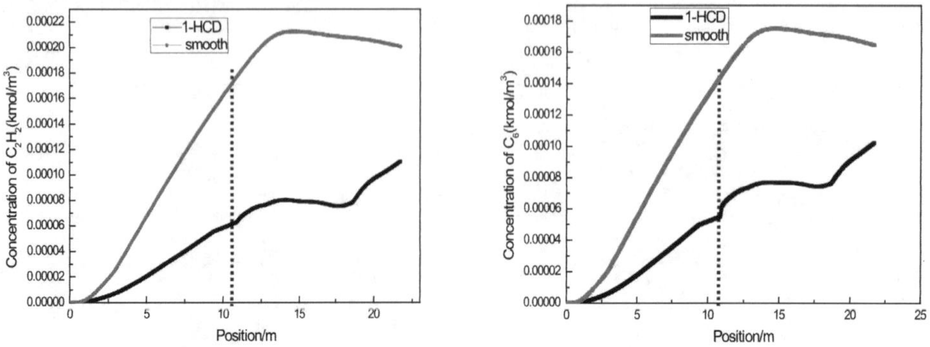

Concentration of coking precursor and heavy component in the wall vicinity

$$R_{coke} = 5.139 \times 10^{11} exp(\frac{-2.0252\times10^{5}}{RT}) \times C_{precursor}^{1.89} \text{[3]}$$

✓ **Significant decrease of coking rate when installing HCD internal.**

[3]P. Kumar, D. Kunzru., The Canadian Journal of chemical engineering, 1985, 63: 598-604

9

Results and discussion

➢Strengthening mechanism:

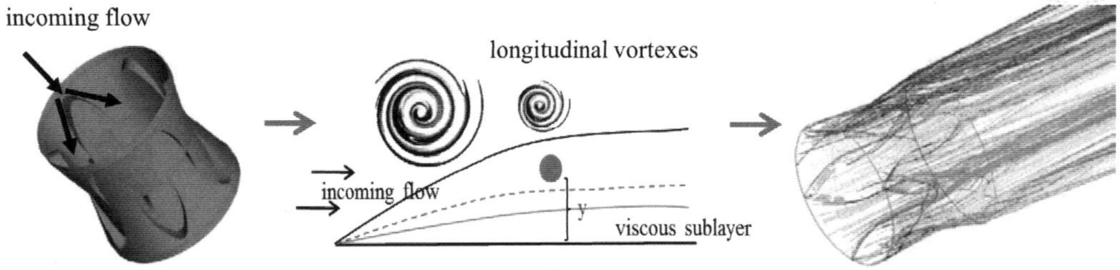

☐ Longitudinal vortex couples disturb boundary layer

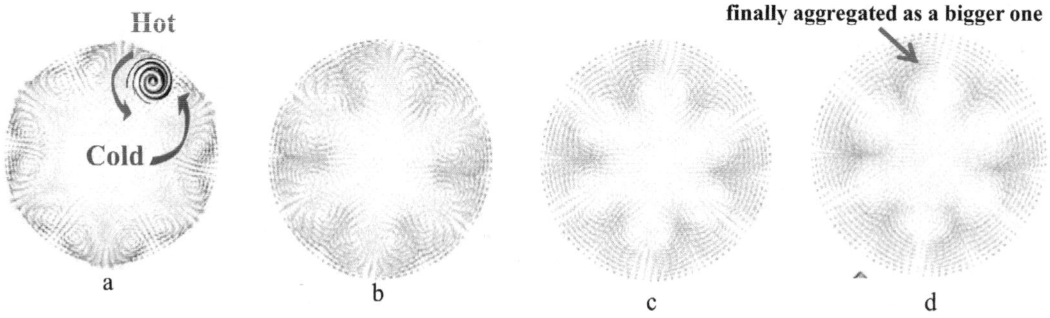

☐ Coalescence of adjacent vortex and gradually moving to the center.

10

Results and discussion

◆ **Heat transfer reinforcement:**

Field synergy theory: $\int \rho c_p (|\vec{U}| \cdot |\Delta T| \cdot \cos\beta) dy = -\lambda \frac{\partial T}{\partial y}$

The bigger the better

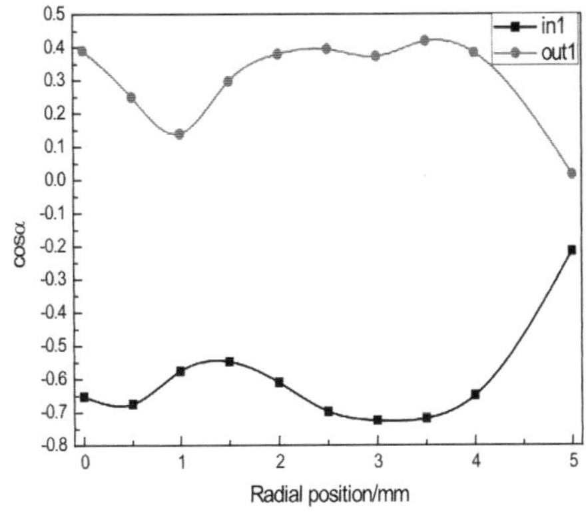

Cosine value of Synergy angle

☐ **Apparent increase of cosβ** at the outlet of HCD internal.

✓ **HCD internal has effectively improved synergetic degree between $\vec{U}$ and ΔT .**

11

Results and discussion

◆ **Mixing enhancement:**

Segregation intensity

$$I = \sum_{i=1}^{n} \frac{(I_i - \bar{I})^2}{n} \rightarrow$$

Case I: 2.2503×10^{-11}

Case II: 1.2437×10^{-11}

Table V: EDC mixing results (mass fraction)

	smooth	HCD
EDC cell volume fraction %	65.75	67.05
Fine scale (C_2H_4)%	0.4607	10.5749
mass fraction(C_2H_4)%	0.4606	10.572
deviation%	0.02171	0.02742
Fine scale mass fraction(C_3H_8)%	69.8003	41.579
mass fraction(C_3H_8)%	69.8007	41.5868
deviation%	5.73E-04	0.01876

Fine scale: $\varepsilon^* = C_\xi (\frac{\nu\varepsilon}{\kappa^2})^{0.75}$

✓ **HCD internal has effectively improved radial homogeneity.**

12

Overall evaluation

➢ Evaluation criteria:

(1) Evaluation method in physical field: **PEC**

Assumption:

- Equal heat transfer surface $A=A_s$
- Constant density with little change in different cases
- Equal resistance wastage $p=p_s$

$$\begin{cases} \frac{K}{K_s} = \frac{hA}{h_sA_s} = \frac{St}{St_s} \cdot \frac{G_s}{G} \cdot \frac{A}{A_s} \\ \frac{p}{p_s} = \frac{f\frac{L}{d}\frac{\rho u^2}{2}Au}{f_s\frac{L}{d}\frac{\rho u_s u^2}{2}A_s u_s} = \frac{f}{f_s} \cdot \left(\frac{G_s}{G}\right)^3 \cdot \frac{A}{A_s} \end{cases}$$

$$\frac{K/K_s}{p/p_s} = \frac{\frac{St}{St_s}\left(\frac{A}{A_s}\right)^{\frac{2}{3}}}{\left(\frac{f}{f_s}\right)^{\frac{1}{3}}}$$ substitute $St = \frac{Nu}{Re \cdot Pr}$

➔ $$\frac{K}{K_s} = \frac{Nu/Nu_0}{(f/f_s)^{\frac{1}{3}}} = PEC$$ **ratio of heat transfer coefficient**

13

Overall evaluation

a. ***Nu/f*** (only heat transfer)　　b. ***Sh/f*** (only mass transfer)

c. Coupled effect of both heat and mass transfer

Heat transfer: Re, Nu, Pr ➔ $St = \frac{Nu}{Re \cdot Pr}$ ➔ $j_H = St \cdot Pr^{\frac{2}{3}}$

Mass transfer: Re, Sh, Sc ➔ $St_m = \frac{Sh}{Re \cdot Sc}$ ➔ $j_D = St_m \cdot Sc^{\frac{2}{3}}$

$j_H = j_D$

➔ $CEC = \frac{j}{f}$

Table VI: Transfer improvement Vs pressure cost

	Re	*Nu*	*F*	*Nu/f*	*Sh*	*Sh/f*	***CEC***
Smooth	5537.2	6.17	0.037	166.75	67.76	1831.23	0.8912
HCD	5569.5	7.08	0.039	181.54	73.38	1881.58	0.9160

✓ HCD internal increases both heat and mass transfer coefficient, but more effective for heat transfer (*Nu/f* increased **8.87%**, *Sh/f* only increased **2.75%**.) 14

Conclusions

1. Olefin yield improvement

Novel coil with HCD internal can effectively improve overall olefin selectivity(**7.12%**) as well as yield(**13.15%**);

2. Coking suppression

Longitudinal vortex couples produced by HCD induce exchanging flow between high temperature zone near the wall and the comparatively cold center zone to release coking.

3. Strengthening mechanism

The strengthening mechanism of cracking process lies in the shear of longitudinal vortex couple to **disturb boundary layer**. And the **radial profile homogeneity** was improved as well.

4. Overall evaluation

Applying CEC to evaluate the combined effect of heat and mass transfer enhancement. Comparing with smooth coil, the CEC increased by 2.78%.

15

East China University of Science and Technology

Thank you for your attention!

428105 Development and Analysis of System-Wide Model for Heterogeneous Catalytic Reactors: A Hierarchical Approach

Abstract:

Heterogeneous catalytic reactors form a large majority of industrial reactors. Catalytic tubular reactors are common in energy and environment applications, where the catalyst is usually coated on the inner surface of the channels. Catalytic monoliths are commonly used as catalytic convertors to reduce hydrocarbon, CO and NOx vehicular emissions, as well as in catalytically stabilized combustion. Catalytic microreactors are being used as thermally integrated reactors for fuel processing and energy applications. With a growing focus on renewable fuels, solid catalyzed systems are being investigated for oil to biodiesel conversion.

Often, the overall system for these applications may consist of multiple reactors in series. The processes are characterized by short time scale, and transients become important. The transport processes include convection of the reacting fluid in the channel, diffusion of the reactants to the catalytic wall, diffusion of the reaction products back to the bulk, heat transfer between the fluid and the wall, and conduction through the solid. The small channel diameters in catalytic monoliths as well as microreactors ensure high catalytic surface area and fast transverse transport. These properties and our extensive knowledge regarding tubular catalytic reactors can be exploited to design efficient catalytic reactors for fuel conversion.

The reactor model is described by partial differential equations in two (or three) spatial dimensions. The complete model is elliptic in spatial domain. Often, computational fluid dynamics (CFD) coupled with appropriate reaction kinetics are used to simulate these systems. Solving such a model is computationally challenging. Hence, several approximations are made the make the problem computationally tractable. For example, a plug flow reactor assumption may be employed for transient simulation at a system level, wherein we assume complete transvers mixing, no axial dispersion and "homogenized" reaction kinetics at bulk fluid conditions. Between these two examples lie other approximations such as transverse transport using Nusselt and Sherwood numbers, boundary layer approximation, combined 1-D and 2-D models, etc.

The objective of this contribution is two-fold. First, we investigate various approximations of the 2D Navier-Stokes simulations to develop a hierarchy of models of decreasing computational complexity. The second aim is to demonstrate the key features of biofuel convertor, and hence provide efficient design and operation strategies thereof. Significant literature exists on various approximations and analysis of heat and mass transfer in catalytic monolith channels. While catalytic convertor and micro-burners are moderate to high temperature gas phase reactors, a biofuel reactor is an example of low tempreature liquid phase reactor. We extend our earlier work on catalytic microreactors to provide a critical analysis of the various assumptions and features of fuel conversion reactors of practical interest.

420527 CFD Investigations of Particle Segregation and Dispersion Mechanisms inside a Polyolefin 8-Leg Loop Reactor of Industrial Scale

Abstract:

The formation of polymer slugs inside loop reactors has long been a troubling issue for the polyolefin industry using slurry–phase process. The mechanism of solid phase segregation occurring in the slurry flow has been reported as one of the root causes of the formation of large polymer slugs. However, the mechanism of solid phase dispersion, which counters the solid phase segregation and retards slug formation, has not been fully understood yet especially from the aspect of fluid dynamics. Therefore, in this study we apply computational fluid dynamic (CFD) simulations to provide insight details about these two competing mechanisms inside an 8–leg loop reactor of industrial scale. The simulations adopt transient Eulerian–Eulerian two fluid model incorporated with the kinetic theory of granular flow to describe the slurry flow consisting of propylene in liquid state and solid polypropylene particles. The slurry flow consisting of solid particles of averaged diameter as 2.5 and 0.5 millimeter are investigated by CFD simulations.

The simulation results indicated that the centrifugal force induced by the bend geometry causes the segregation of the solid particles inside the bottom bends, resulting in thick particulate ropes close to the outer pipe wall. The segregation extents are different depending on the location of the bends: no clear particulate layers are found inside the top bends or the bends close to the axial flow pump. As the slurry flow is forced to change its axial flow direction inside the 2nd and 6th bends, the sudden change of flow directions induces "double–bend" effect. Governed by such effect, the secondary flow on the cross sections of vertical legs exhibits the "single–vortex" structure. This type of flow structure enhances the solid dispersion mechanism inside vertical legs. While rotating slowly inside vertical legs, particulate ropes are gradually dispersed and mixed with the liquid phase by the secondary flow as well as the turbulent dispersion. In the initial stage prior to the formation of large slugs, the competition between the segregation and dispersion mechanisms generates multiple slurry clusters of various solid volume fractions. When these clusters are circulated inside the loop reactor by the axial flow pump, the monitors on the cross sections of the loop reactor observe fluctuating profiles of the operating parameters such as the solid volume fraction and velocity magnitudes. When these slurry clusters pass through the pump region, fluctuating pump pressure output, in other words pump power consumption, is observed due to the varying frictional force exerted to the pump impeller. For example, if clusters containing large amounts of solid particles pass through the pump impeller, the frequent interactions between the solid particles and pump impeller generates strong frictional force. In order to maintain its constant rotational speed, the pump has to consume significant amount of power to overcome the friction force. In contrast, the pump requires less amount of power to drive the flow when clusters having smaller amount of particles present in the impeller region. If the solid segregation mechanism is not controlled properly, the particulate ropes can eventually develop into large slugs as the segregation phenomenon proceeds for 6 to 8 hours. The movements of large slugs inside loop reactors cause rapid and violent pump power fluctuations that the safety interlocks has to shut down the entire production process.

The simulation results suggested that the solid segregation can be suppressed effectively if particle sizes are reduced. Small particles do not stratify into thick particulate ropes inside bottom bends, and they are distributed more uniformly in loop reactors compared to large particles. Although reducing particle sizes can retard the formation of large slugs, such technique is challenging to adopt as particle sizes are generally determined by reaction conditions. Other mitigation techniques, *i.e.*, using bypass line to loop reactors, are more feasible to control the formation of large slugs in the loop reactor.

Semi Analytical Solution of a Heat Transfer and Kinetic Models Applied in a Biomass Pyrolysis Reactor

* *S.L. Bertoli, J. de Almeida Jr*[1], *J.C.S.C. Bastos, M. K. da Silva, G.J. Pacheco, V.R. Wiggers*

**savio@furb.br*

Abstract

Currently, fuels renewal and innovation has been intensively investigated by scientific community. Nevertheless, thermal cracking (pyrolysis) is considered a promising technique in biofuels production. This paper proposes a semi analytical solution of a mathematical model for heat transfer and chemical conversion phenomena applied to a biomass pyrolysis reactor. Equations set were solved through the Finite Analytical Method. Thus, it was possible to incorporate the non-linear terms in the solution and develop a numerical procedure in which time interval can be made arbitrarily small without stability and convergence problems. The thermochemical process is a decomposition reaction (reduction) occurring at higher temperatures, in an environment quasi or deoxygenated – a rupture of the original molecular structure of a given compound or mixture by heating. This is self-sustaining from an energy standpoint, since the 'break' in a waste processing reactor produces surplus energy. The biomass pyrolysis reactor was simulated, based on works of heat and mass transfer models and solved by Walas [1959] and modified by Leung et al. [1965] and also in the studies of Bertoli [2000], Meier et al. [2009] and Wiggers et al. [2009]. Steady-state, uniform average temperature, adiabatic reactor, inert particles, first-order reactions were the simplifying hypothesis of the mathematical model. The semi analytical solution was realized considering the nonlinearity of the equations system by partitioning the reactor in finite intervals; the linear terms remains and the non-linear are made constant in the intervals. Finite Analytical Method is based on the numerical solution incorporating a local solution of the partial or ordinary differential equations. For data comparison (semi analytical versus numerical solutions) Runge-Kutta-Fehlberg routine was used. The proposed method was successfully applied and the results between semi analytical and numerical solutions agree with minimum error values.

Keywords: *Semi analytical solution, Pyrolysis, Heat transfer model, Kinetic model, Biomass reactor*

Introduction

A few decades 'specialists' work on biofuels an alternative for fossil fuels. However, due to resilience marketing, this new fuel must adapt an existing technology.

A wide range of processes, equipment and materials have been studied over time. Since 1973, with the energy crisis, pyrolysis has become thoroughly studied. However, there is a strong concern that such energy sources are not only renewable, but can also impact as little as possible on the environment.

Derived from the Greek word pyrolysis means 'pyr = pyros' (fire) and 'lysis' (break), and is defined as the irreversible thermal decomposition of organic compounds in the oxygen absence [TSAI et al., 2009].

Phenomenological studies of pyrolysis processes are the key for understanding and analysis, enabling optimization and scale-up [MEIER et al., 2014; WIGGERS, 2009].

For this intention, a mathematical model and their respective solution technique of a pyrolysis reactor is proposed, as follow.

Mathematical model

Moving bed reactor is a continuous vessel used in several industrial processes involving fuels [FOGLER, 2009].

Heat transfer modeling and chemical kinetics are of critical importance in reactional process; existing several studies about them [GOLDSTEIN, 2006; 2010a-b].

Models used in this work are similar to those found in literature, such as Walas [1959], modified by Leung et al. [1965] and also in the Bertoli [2000], Meier et al. [2009] and Wiggers et al. [2009]. Figure 1 represents a moving bed reactor:

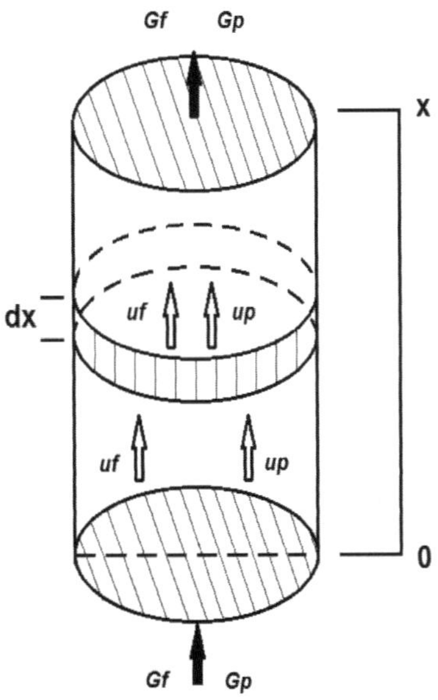

Figure 1 – Reactor scheme

G_f is the gas molar flow (kmol.m^{-2}.s^{-1}), W_p is the solid mass flow (kg.m^{-2}.s^{-1}), u_f is the superficial fluid velocity (m.s^{-1}), u_p is the particle velocity (m.s^{-1}) and dx is a differential distance in the axial direction of the control volume.

Main assumptions made in this analysis are similar those described in [BABU et al., 2002], [BERTOLI, 1989; 2000], [MEIER et al., 2009] and [BERTOLI et al., 2012]:

- Steady state
- Uniform particle and fluid temperature
- Adiabatic reactor
- Inert and spherical particles
- First order chemical reaction
- Flow developed in the two phase system
- Fluid and particle uniform velocity in cross sections

The model equations used are described below.

Fluid Temperature

Using the time-residence concept ($t = x/u_p$), Equations 1-3 describe the fluid temperature behavior and can be divided into two contributions:

First one, $\beta_1(T_p-T_f)$, deals with the fluid-particle convection heat transfer.

Second contribution, $\beta_2(f_n)$, is a predominantly endothermic phenomenon originated from thermal cracking [BAUM et al., 1974]. Analogous modeling is found in Walas [1959].

$$\frac{dT_f}{dt} = \beta_1(T_p - T_f) + \beta f_N \tag{1}$$

$$\beta_1 = \frac{u_p.A.h_p}{G_f.M.c_f} \tag{2}$$

$$\beta = \frac{\varepsilon.k.P.(-\Delta H_T)}{\varkappa.\rho_f.c_f.(1+\epsilon.x).R.T_f} \tag{3}$$

h_p is the convective heat transfer coefficient of the gas-solid interaction ($W.m^{-1}.°C^{-1}$), A is the external particle surface area per unit reactor volume ($m^2.m^{-3}$), M is the average molecular weight of the fluid ($kg.kmol^{-1}$), κ is the ratio of the fluid velocity to the particle velocity, c_f is the fluid heat capacity ($W.kg^{-1}.°C^{-1}$), ε is the bed porosity, ϵ is the fractional change in volume per reacted mol, k is the rate constant (s^{-1}), P is the thermodynamic pressure (Pa), ΔH_T is the reaction heat ($J.kmol^{-1}$), x is the conversion, ρ_f is the fluid density ($kg.m^{-3}$) and R is the ideal gas constant ($J.kmol^{-1}.K^{-1}$). T_p and T_f are the particle and fluid temperature (K), respectively.

Particle Temperature

Equation 4 is used to describe the particle temperature over time, similar to models found in Bertoli et al. [2012].

$$\frac{dT_P}{dt} = \beta_2(T_f - T_p) \tag{4}$$

$$\beta_2 = \frac{A.h_p}{\rho_p.c_p.(1-\varepsilon)} \tag{5}$$

ρ_p is the particle density ($kg.m^{-3}$) and c_p is the particle heat capacity ($W.kg^{-1}.°C^{-1}$)

Conversion Component

Equation 6 describes the reagent conversion over time.

$$\frac{dF_n}{dt} = -\alpha f_n \tag{6}$$

$$\alpha = \frac{u_p.\varepsilon.k.P_{l-1}}{(1+\epsilon.x_{l-1}).R.T_f.G_f} \tag{7}$$

$$F_n = 1 - x; u_p = \frac{W_p}{(1-\varepsilon)}; k = 10^{\left(a - \frac{b}{T_{f\,l-1}}\right)} \tag{8}$$

a and b are the coefficients of the experimental reaction rate [Leung et al. 1965].

For the model equation, Runge-Kutta-Fehlberg routine types were used to compare the obtained results with the semi analytical solution.

Approaching Method

Semi analytical solution considers the nonlinearity of the equation system by partitioning the reactor in finite intervals, in which non-linear terms are kept constant and evaluated under the temperature conditions at the beginning of the interval.

The Finite Analytical Method (FAM) is based on numerical solution incorporating local solutions of partial and ordinary differential equations. The FAM decomposes the overall region of a problem governed by differential equations in small elements in which local analytical solutions are obtained due to local geometry simplicity, equations and boundary conditions [Lin, 1996].

Rearranging Equations 1 and 4, separating the variables and simplifying:

$$\frac{dT_f}{dt} = \frac{\beta_1}{\beta_2}\frac{dTp}{dt} + \beta f_n \quad (9)$$

$$\frac{dT_p}{dt} + \beta_2 T_p = \beta_2 T_f \quad (10)$$

$$\frac{\left({}^{df_n}\!/_{f_n}\right)}{dt} = -\alpha \quad (11)$$

Rewriting Equations 9 to 11, in order to obtain the corresponding differentials:

$$dT_f = -\frac{\beta_1}{\beta_2} dT_p + \beta f_n dt \quad (12)$$

$$d(e^{\beta_2 t} T_p) = e^{\beta_2 t} \beta_2 T_f dt \quad (13)$$

$$d \ln f_n = -\alpha dt \quad (14)$$

Integrating Equations 12 to 14, in the interval t_l to t_{l-1} and rearranging them:

$$T_{fl} = T_{fl-1} - \frac{\beta_1}{\beta_2}\left(T_{pl} - T_{p_{l-1}}\right) + \int_{t_{l-1}}^{t_l} \beta f_n dt \quad (15)$$

$$T_{pl} = T_{pl-1}\, e^{-\beta_2 dt} + \beta_1 e^{-\beta_1 t_l} \int_{t_{l-1}}^{t_l} e^{\beta_2 t_l}\, T_f dt \quad (16)$$

$$f_n = f_{n_{l-1}} e^{-\int_{t_{l-1}}^{t_l} \alpha dt} \quad (17)$$

Equations 18 to 20 present the recursion expression from the FAM:

$$\int_{t_{l-1}}^{t_l} \alpha dt \cong -\alpha_{l-1}\Delta t \qquad => \qquad f_n \cong f_{n_{l-1}} e^{-\alpha_{l-1}(t-t_{l-1})} \tag{18}$$

$$\int_{t_{l-1}}^{t_l} \beta f_n \, dt \cong \beta_{l-1} f_{n_{l-1}} e^{\alpha_{l-1}t_{l-1}} \int_{t_{l-1}}^{t_l} e^{-\alpha_{l-1}t} dt \cong -\frac{\beta_{l-1}}{\alpha_{l-1}} f_{n\,l-1}(e^{-\alpha_{l-1}\Delta t} - 1) \tag{19}$$

$$\int_{t_{l-1}}^{t_l} e^{\beta_2 t} T_f \, dt \cong \frac{T_{f_{l-1}}}{\beta_1} e^{\beta_2 T}\left(1 - e_l^{-\beta_2 \Delta t}\right) \tag{20}$$

Substituting Equations 18 to 20 in Equations 15 to 17 results (Equations 21 to 23):

$$T_{fl} = T_{fl-1} - \frac{\beta_1}{\beta_2}(T_{pl} - T_{p_{l-1}}) + \left(\frac{\beta_{l-1}}{\alpha_{l-1}}\right) f_{n_{l-1}}(1 - e^{-\alpha_{l-1}\Delta t}) \tag{21}$$

$$T_{pl} = T_{pl-1}\, e^{-\beta_2 \Delta t} + \frac{T_{f_{l-1}}}{\beta_2}(1 - e^{\beta_2 \Delta t}) \tag{22}$$

$$f_n = f_{n_{l-1}} e^{-\alpha_{l-1}\Delta t} \tag{23}$$

Equations 21 to 23 are rewrites as:

$$T_{fl} - T_{fl-1} = -\frac{\beta_1}{\beta_2}(T_{pl} - T_{p_{l-1}}) - \left(\frac{\beta_{l-1}}{\alpha_{l-1}}\right) f_{n_{l-1}} e^{\alpha_{l-1}t_{l-1}} \,(e^{-\alpha_{l-1}t_l} - e^{\alpha_{l-1}t_{l-1}}) \tag{24}$$

$$e^{\beta_2 t_l} T_{pl} - T_{pl-1}\, e^{\beta_2 t_{l-1}} = \frac{T_{fl-1}}{\beta_2}(e^{\beta_2 t_l} - e^{\beta_2 t_{l-1}}) \tag{25}$$

$$\ln f_{n_l} - \ln f_{n_{l-1}} = -\alpha_{l-1}\Delta t \tag{26}$$

Using the difference operator Δ, the Equations 24 to 26 assumes the following form:

$$(\Delta T_f)_l = -\frac{\beta_1}{\beta_2}(\Delta T_p)_l - \left(\frac{\beta_{l-1}}{\alpha_{l-1}}\right) f_{n_{l-1}} e^{\alpha_{l-1}t_{l-1}} \left(\Delta e^{-\alpha_{l-1}t}\right)_l \tag{27}$$

$$(\Delta e^{\beta_2 t_l} T_p)_l = \frac{T_{fl-1}}{\beta_2}\left(\Delta e^{\beta_2 t}\right)_l \tag{28}$$

$$(\Delta \ln f_n)_l = -\alpha_{l-1}\Delta t \tag{29}$$

Summing up in l from $l = 0$ to $l = \text{n}$

$$T_{fn} - T_{fl} = -\frac{\beta_1}{\beta_2}\left(T_{pN} - T_{po}\right) - \sum_{l=1}^{n}\left(\frac{\beta_{l-1}}{\alpha_{l-1}}\right) f_{n_{l-1}} e^{\alpha_{l-1} t_{l-1}} (\Delta e^{-\alpha_{l-1} t})_l \tag{30}$$

$$T_{pN} e^{\beta_2 t_n} - T_{po} = \frac{1}{\beta_2} \sum_{l=1}^{N} T_{f_{l-1}} \left(\Delta e^{\beta_2 t}\right)_l \tag{31}$$

$$\ln f_n - \ln f_0 = -\sum_{l=1}^{n} (\alpha_{l-1} \Delta t)_l \tag{32}$$

Making $N \to \infty$:

$$T_f - T_{fo} = -\frac{\beta_1}{\beta_2}\left(T_p - T_{p_o}\right) - \int_0^t \frac{\beta}{\alpha} f_n e^{\alpha t} \, d(e^{-\alpha t}) \tag{33}$$

$$T_p = e^{-\beta_2 t} T_{po} + \frac{e^{-\beta_2 t}}{\beta_2} \int_0^t T_f \, d(e^{\beta_2 t}) \tag{34}$$

$$\ln f_n = -\int_0^t \alpha dt \tag{35}$$

Finally, developing the differentials in Equations 33 to 35 yields the model exact solution (formal):

$$T_f = T_{fo} - \frac{\beta_1}{\beta_2}\left(T_p - T_{p_o}\right) + \int_0^t \beta f_n \, dt \tag{36}$$

$$T_p = e^{-\beta_2 t} T_{po} + e^{-\beta_2 t} \int_0^t T_f \, e^{\beta_2 t} dt \tag{37}$$

$$f_n = -e^{-\int_0^t \alpha dt} \tag{38}$$

In this way is proved the convergence of the method to the analytical solution. In the following, after semi analytical solution development, results are compared with RKF method.

Results

Since the analytical solution of this case is a difficult problem, the results were then compared with a numerical consolidated solution. However, in order to obtain the magnitude of possible numerical errors, it had also been compared in Figure 2 the semi analytical results *versus* no chemical reaction analytical solution, where the results match almost perfectly.

Figures 3 and 4 present the semi analytical and RKF solutions of the model (Eqs 1, 4 and 6) for hydrocarbon cracking ($C_4H_{10} \rightarrow 2C_2H_4 + H_2$) [WALAS, 1959] that is a well know problem and retains also major characteristics of a gas-phase pyrolysis process and gas-phase pyrolysis of biomass process [WIGGERS et al. 2009], where it is not possible to observe disparity between the solutions. In the

Figure 3 time-concentration profiles are showed, where can be noted an initial induction period (approx. 0,3s) and afterwards the concentration increase exponentially until the maximum conversion be attained. The error between the solutions is minimal (10^{-6} order) as can be seen.

In the Figure 4 time-temperature profiles for both phases are presented. The error is minimal (10^{-6} order) too. Initially, there is a steep increase in the fluid temperature and a slight decrease of the solid temperature, which is due to the small thermal capacitance rate ratio $[(G_f.M.c_f)/(W_p.c_p) << 1]$ and the fact that during the induction period there is no heat source. After approximately 1s, it is initiated the endothermic chemical reaction, slowing the heating rate of the fluid.

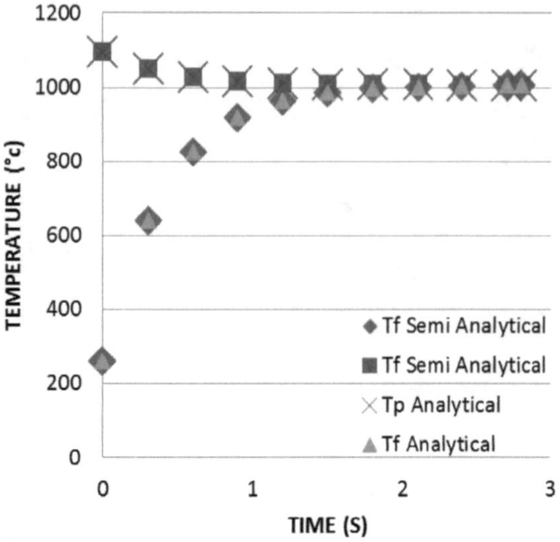

Figure 2 – Time-temperature profiles (No Chemical reaction)

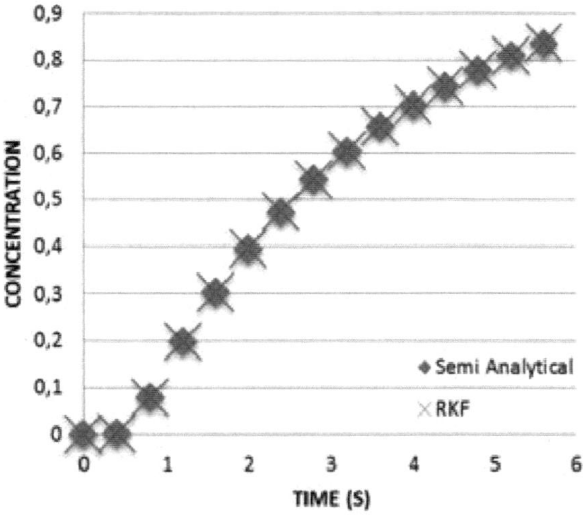

Figure 3 – Time-concentration profiles.

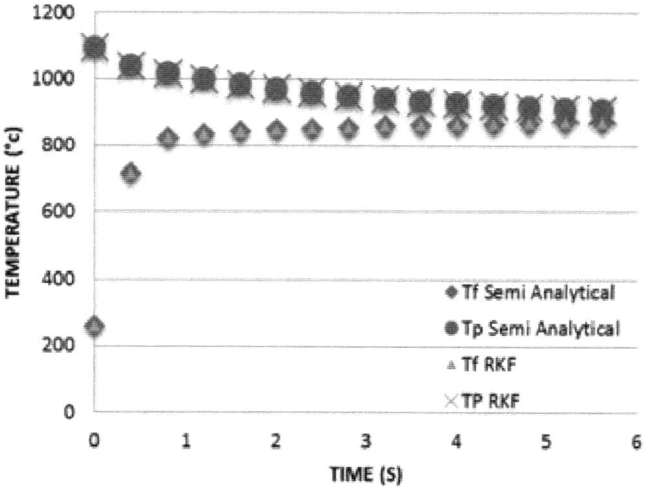

Figure 4 – Time-temperature profiles.

Conclusions

Models solution enables pyrolysis reactors scale-up, which can be applied for biofuels and subproducts manufacture.

FAM allowed the ordinary differential equations discretization providing a way to improve stability and quality of the solution.

For the case studied, semi analytical solution was effective with excellent accuracy in the solution of nonlinear ODE's system for thermal cracking, which suggests their application in other systems of nonlinear ODE's.

References

1. Walas, S. M. (1959), Reaction kinetics for chemical engineers. McGraw-Hill Book Co. New York.

2. Leung, P. K., Quon D. (1965), A Computer Model for Moving Beds – Chemical Reaction in Fluid Phase Only, The Canadian Journal of Chemical Engineering.

3. Bertoli S.L. (2000), Radiant and convective heat transfer on pneumatic transport of particles: an analytical study International Journal of Heat and Mass Transfer, 43, 2345-2363.

4. Meier H. F., Noriler D., Bertoli S. L. (2009), A solution for a heat transfer model in a moving bed through the self-adjoint operator method, Latin American Applied Research, 327-336.

5. Wiggers, V. R. (2009), Biofuels from continuous fast pyrolysis of soybean oil: A pilot plant study. Bioresource Technology, 100, 6570-6577.

6. Tsai, W., Chang, J., Hsien, K., Chang Y. (2009), Production of Pyrolytic Liquids from Industrial Sewage Sludges in an Induction-Heating Reactor. Bioresource Technology, 100.

7. Meier H. F., Wiggers V.R., Zonta G.R., Scharf D.R., Simionatto E.L., Ender L. (2014), A kinetic model for thermal cracking of waste cooking oil based on chemical lumps, The Science and Technology of fuel and Energy.

8. Fogler, H. S. (1999), Elements of Chemical Reaction Engineering. 3°ed. New Delhi: Prentice-Hall.

9. Goldstein R.J.; Ibele W.E.; Patankar S.V.; Simon T.W.; Kuehn T.H.; Strykowski P.J.; Tamma K.K.; Heberlein J.V.R.; Davidson J.H.; Bischof J.; Kulacki F.A.; Kortshagen U.; Garrick S.; Srinivasan V. Ghosh K.; Mittal R. (2010), Heat transfer—A review of 2004 literature. International Journal of Heat and Mass Transfer, 53, 4343–4396.

10. Goldstein R.J.; Ibele W.E.; Patankar S.V.; Simon T.W.; Kuehn T.H.; Strykowski P.J.; Tamma K.K.; Heberlein J.V.R.; Davidson J.H.; Bischof J.; Kulacki F.A.; Kortshagen U.; Garrick S.; Srinivasan V. Ghosh K.; Mittal R. (2010), Heat transfer - A review of 2005 literature. International Journal of Heat and Mass Transfer, 53, 4397-4447.

11. Goldstein R.J.; Ibele W.E.; Patankar S.V.; Simon T.W.; Kuehn T.H.; Strykowski P.J.; Tamma K.K.; Heberlein J.V.R.; Davidson J.H.; Bischof J.; Kulacki F.A.; Kortshagen U.; Garrick S.; Srinivasan V. (2006), Heat transfer - A review of 2003 literature. International Journal of Heat and Mass Transfer, 49, 451–534.

12. Babu, B. V.; Chaurasia, A. S. (2002), Modeling & Simulation of Pyrolysis: Effect of Convective Heat Transfer & Order of Reactions. Birla Institute of Technology & Science.

13. Bertoli S. L., (1989), Radiant and Convective Heat Transfer in Moving Beds, M. Sc. Thesis, COPPE, Rio de Janeiro, Brazil (in Portuguese).

14. Bertoli, S. L., Valle, J. A. B., Gerent, A. G., Almeida, J. (2012), Heat transfer at pneumatic particle transport - Limit solutions. Powder Technology, v. 232, p. 64–77.

15. Baum, B.; Parker C. H. (1974), Solid Waste Disposal: Reuse, Recycle and Pyrolysis Arbor Science Publishers Inc.

16. Lin, W., Haik, Y., Bernatz, R., Chen C. J. (1996) Finite analytic method and its applications: a review. Dynamics of Atmospheres and Oceans.

421867 A Multi-Scale Two-Dimensional Packed Bed Reactor Model for Catalytic Steam Methane Reforming

Abstract:

Steam methane reforming (SMR) is the primary source of synthesis gas (CO/H_2) production for use in direct reduction of iron-ore. The endothermic and equilibrium-limited SMR reaction is carried out in multiple tubes packed with reforming catalyst. The models available in the literature for SMR packed bed reactor are one dimensional homogeneous and pseudo-homogeneous. The homogeneous model calculates the reaction at bulk fluid concentration and pseudo-homogeneous employs catalyst effectiveness factor. These models do not take into account the reaction and diffusion in the catalyst particle, and therefore predict inaccurate conversion and selectivity of products.

This talk presents a steady state 2D multi-scale packed bed reformer model that takes into account the solid-gas phase mass and energy diffusional limitations along the reactor length. In lieu of using the catalyst effectiveness factor, mass and energy balance in the 3D catalyst particle is calculated and coupled with gas phase model. The gas phase model takes into account the axial and radial dispersion as well as finite solid-fluid mass and heat transfer coefficients; calculated using empirical correlations available in the literature. The pressure drop in the packed bed is calculated employing the Ergun equation and compared with the Eisfeld-Schnitzlein equation that takes into account the wall effects. The molar expansion during reaction is also taken into account using a differential equation for average molecular weight along reactor length.

This talk presents the simulation results for different case studies employing the model. First, we will present the 2D non-isothermal model results and compare with 1D model results. Second, we will compare the simulation results using two different pressure drop equations. Finally, we will compare the results without taking into account the mole effects on velocity during reaction.

431069 Kinetics of Isothermal and Non-Isothermal Fast Pyrolysis of Alkali Lignin

Thursday, November 12, 2015: 8:30 AM
250B (Salt Palace Convention Center)

Deepak Ojha, Chemical Engineering, Indian Institute of Technology Madras, Chennai, India, Chennai, TN, India and R. Vinu, Department of Chemical Engineering, Indian Institute of Technology Madras, Chennai, India, Chennai, TN, India

Kinetics of Isothermal and Non-isothermal Fast Pyrolysis of Alkali Lignin

Deepak Kumar Ojha and R. Vinu

Department of Chemical Engineering and National Center for Combustion Research and Development, Indian Institute of Technology Madras, Chennai- 600036, India

E-mail: deepakojha86@gmail.com, vinu@iitm.ac.in

Abstract

Fast pyrolysis has received enormous attention for the conversion of biomass and waste plastics into usable intermediates and valuable chemicals. The involvement of hundreds of chemical reactions makes the process extremely complex. The operating parameters such as pyrolysis temperature and vapor residence time play a decisive role in determining the product spectrum. High temperatures lead to enhanced conversion but also promote the formation of condensed ring aromatics that can be undesired for various reasons, while long vapor residence times lead to more vapor phase interactions, thereby promoting secondary reactions that will often lead to more char and undesired product formation (<C4 oxygenates and non-condensable gases). The knowledge of apparent kinetic parameters, viz. activation energy and frequency factor, and optimal reaction time can be very useful in designing a sophisticated fast pyrolysis reactor.

The objectives of this work are two fold: firstly, the apparent kinetic parameters for isothermal and non-isothermal fast pyrolysis of alkali lignin were evaluated. Conventionally, the kinetic parameters are reported only for slow or medium heating rate pyrolysis via thermogravimetry experiments. However, in this work, kinetics of fast pyrolysis is determined at heating rates greater than 1000 °C/s. Secondly, the time evolution of various organic functional groups formed in the vapor phase during fast pyrolysis of alkali lignin was studied. The fast pyrolysis experiments are conducted in a Pyroprobe® 5200 pyrolyzer (CDS Analytical Inc.) and the generated vapors were characterized by Fourier transform infrared spectrometer (FT-IR, Cary 660, Agilent Technologies) which is equipped with a high sensitivity MCT detector. The condensable and non-condensable products were thoroughly characterized using 2D-GC/MS (Agilent Technologies) and a gas analyzer (Bhoomi Analyzers, India).

Figure 1 depicts the FT-IR spectra of the major functional groups evolved during fast pyrolysis of alkali lignin at 500 °C. The peak at 3585 cm^{-1} indicates the formation of non-hydrogen-bonded phenolic and alcoholic moieties. Interestingly alkali lignin does not exhibit significant vibration above 3500 cm^{-1} but it shows a vibration below 3500 cm^{-1} indicating that most of the OH groups present in lignin are hydrogen bonded in lignin matrix. The aliphatic vibrations in the pyrolysis vapor are because of C-H bonds present in propyl chain of lignin matrix whereas aromatic vibrations are as a result of vibration of C-H bonds in benzene ring. The intensity of C-H aliphatic bonds was predominant over C-H aromatic vibration at moderate temperatures (400 °C). But with the increase in pyrolysis temperature, the intensity of the C-H aromatic vibration was found to dominate the C-H aliphatic vibration. This is because of the cleavage of propyl subunits of lignin into the smaller fragment such as CH_4, CO and CO_2. The formation of these non condensable gases was confirmed by FT-IR spectra and gas analyzer. The time required for maximum production of products at 400 °C was 12 s and it reduced to 8 s at 800 °C. The first order apparent activation energies for isothermal and non-isothermal degradation were calculated to be 14.29 kJ/mol and 25.53 kJ/mol, respectively.

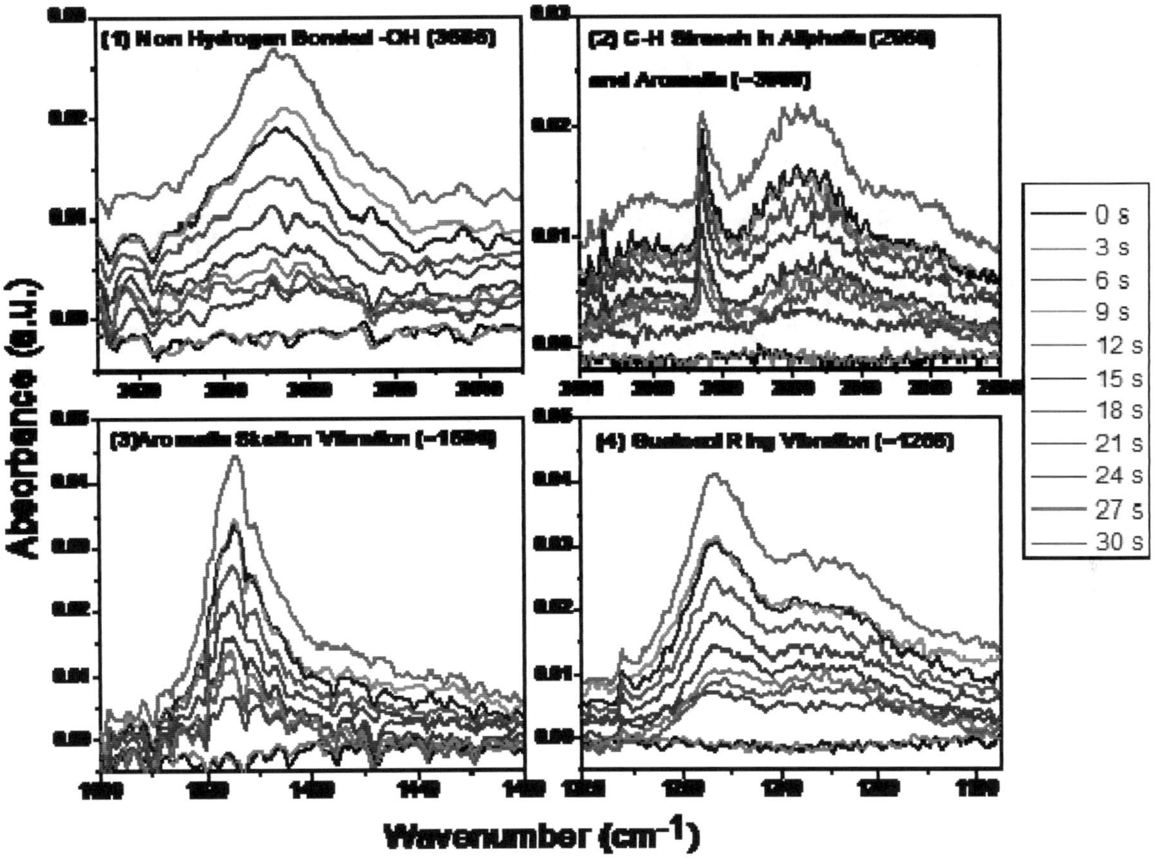

Figure 1: Time evolution of major functional groups during fast pyrolysis of alkali lignin at 500 °C.

1

UNIVERSIDAD NACIONAL DE COLOMBIA SEDE MANIZALES

15AIChE Annual Meeting, Salt Lake City, UT

Cenicafé

Comparison of the synthesis gas production via gasification and combustion from Coffee Cut-Stems

CARLOS A. GARCÍA[A], CARLOS O. TASCON[B], CARLOS A. CARDONA[A]

[A] INSTITUTO DE BIOTECNOLOGÍA Y AGROINDUSTRIA, DEPARTAMENTO DE INGENIERÍA QUÍMICA. UNIVERSIDAD NACIONAL DE COLOMBIA SEDE MANIZALES.

[B] NATIONAL COFFEE RESEARCH CENTER . CENICAFÉ. CHINCHINÁ, CALDAS, COLOMBIA.

Outline

2

Agricultural Residues

1. Introduction 3

Agricultural Crops

Harvesting

Crop Residues

Processing

Agroindustrial Residues

Coffee Cut-Stems

1. Introduction 4

Colombia is the fourth largest producer of Coffee in the World.

80,000 hectares of Coffee tree Wood are renovated

17 tonne of dry Wood per day can be obtained

690 GWe can be generated

Heating and Cooking

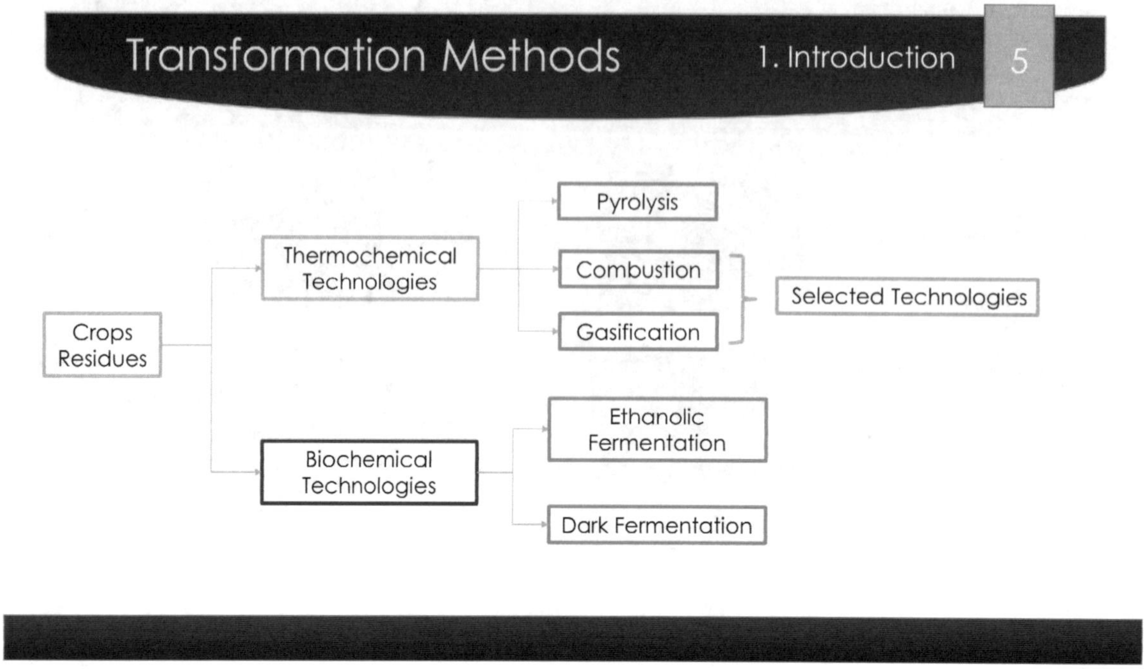
Transformation Methods
1. Introduction
5
Crops Residues
Thermochemical Technologies
Biochemical Technologies
Pyrolysis
Combustion
Gasification
Selected Technologies
Ethanolic Fermentation
Dark Fermentation

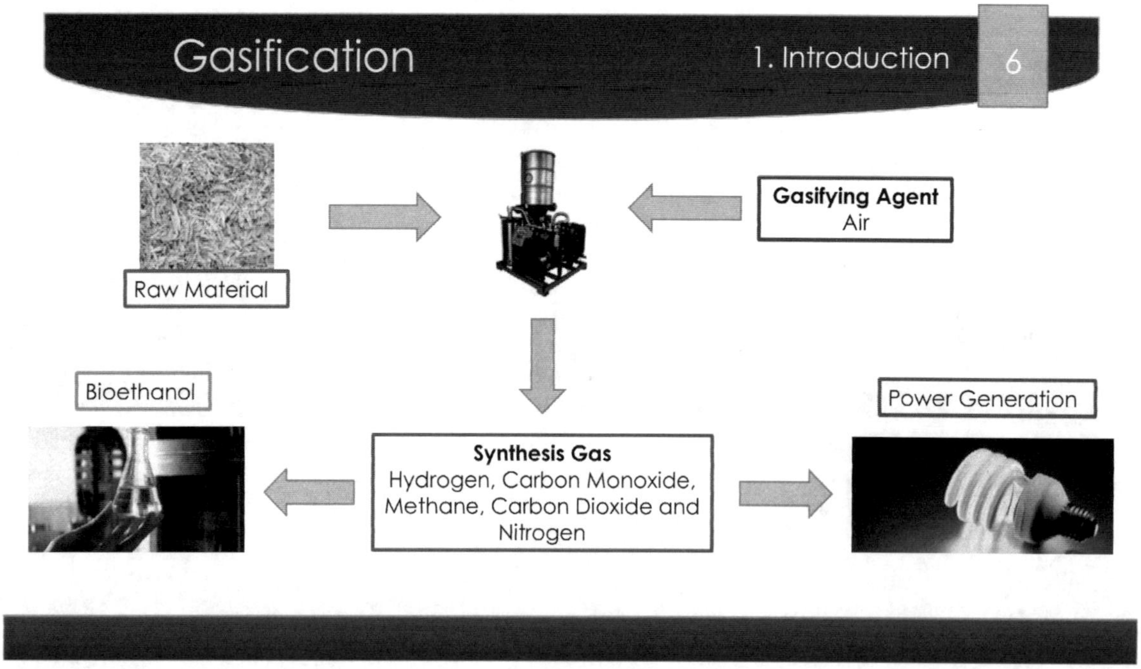
Gasification
1. Introduction
6
Raw Material
Gasifying Agent
Air
Synthesis Gas
Hydrogen, Carbon Monoxide, Methane, Carbon Dioxide and Nitrogen
Bioethanol
Power Generation

Combustion

1. Introduction 7

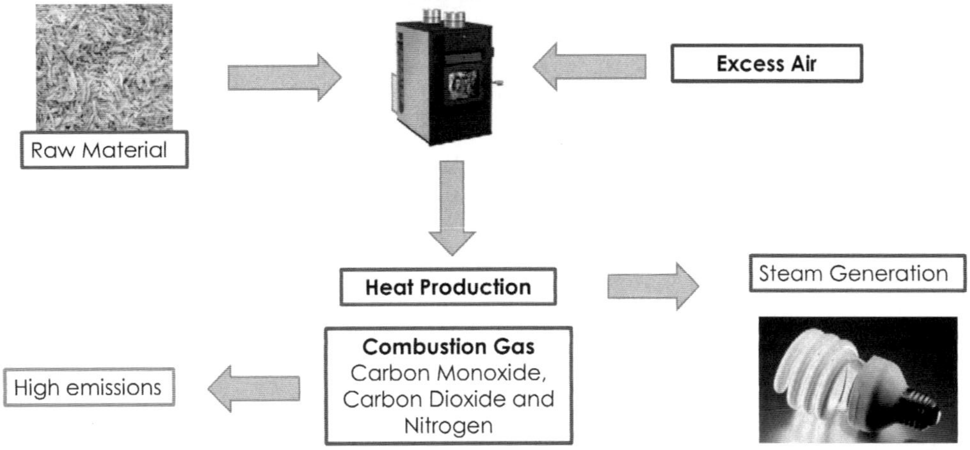

Objective

8

- The aim of this study is to evaluate and compare the energy and environmental assessment of two processes for energy production (combustion and gasification) from Coffee Cut-Stems, with particular emphasis on the gasification process as an alternative to approach the integral lignocellulosic residues with high energy potential and low emissions.

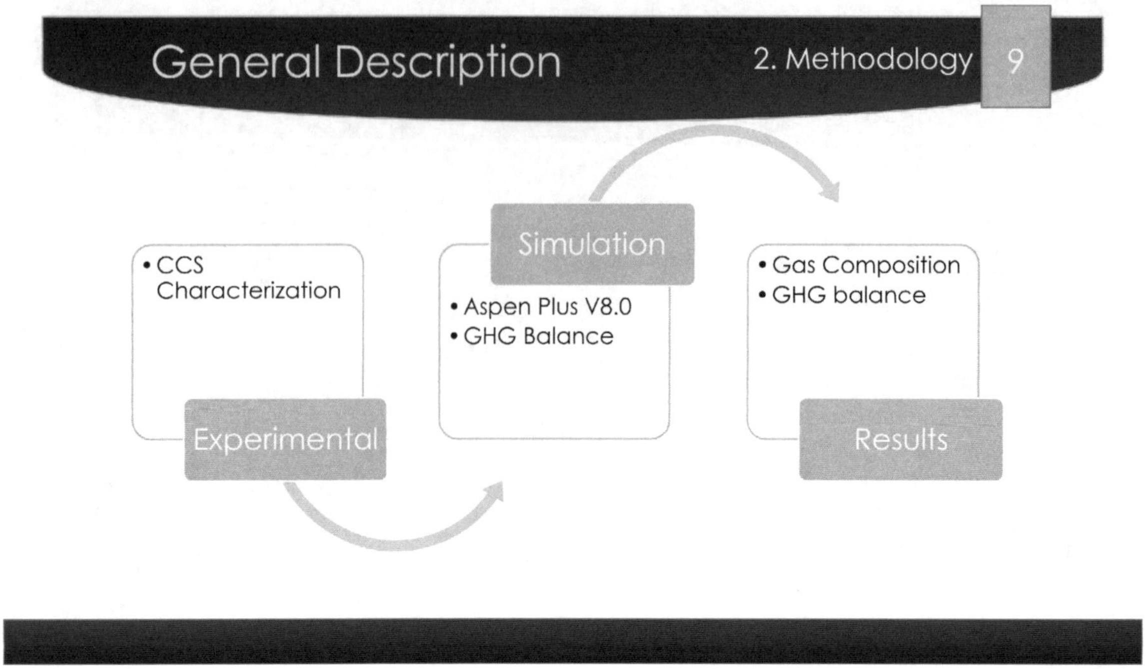
General Description
2. Methodology
9
CCS Characterization
Experimental
Simulation
Aspen Plus V8.0
GHG Balance
Gas Composition
GHG balance
Results

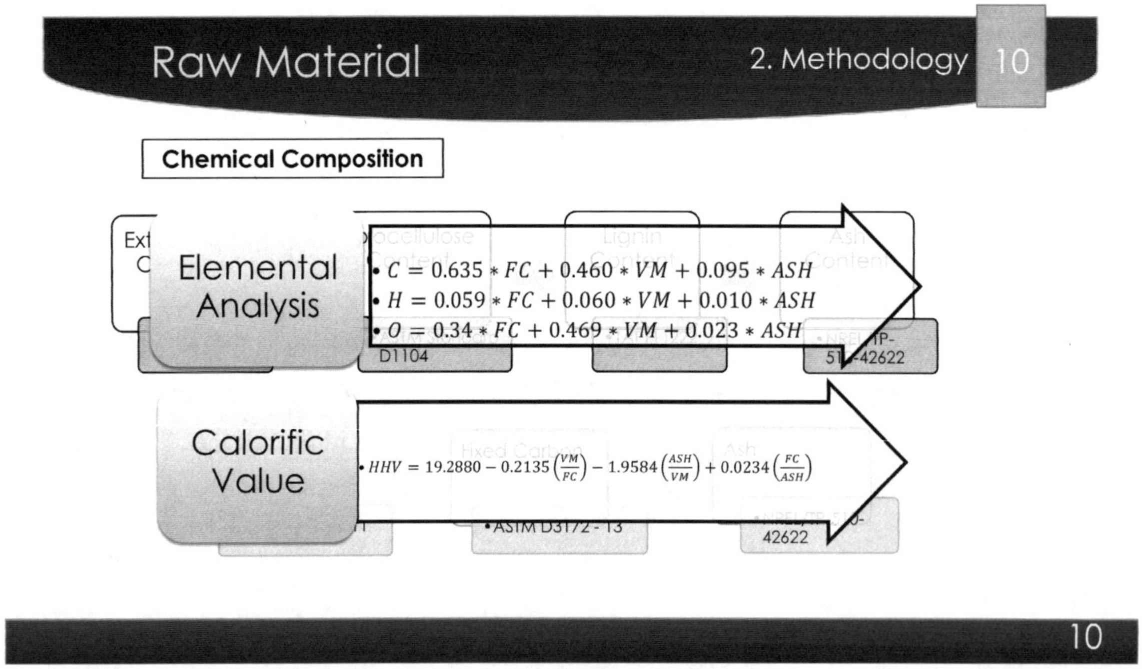
Raw Material
2. Methodology
10
Chemical Composition
Elemental Analysis
$C = 0.635 * FC + 0.460 * VM + 0.095 * ASH$
$H = 0.059 * FC + 0.060 * VM + 0.010 * ASH$
$O = 0.34 * FC + 0.469 * VM + 0.023 * ASH$
D1104
Calorific Value
$HHV = 19.2880 - 0.2135\left(\frac{VM}{FC}\right) - 1.9584\left(\frac{ASH}{VM}\right) + 0.0234\left(\frac{FC}{ASH}\right)$
42622
10

Simulation Procedure

2. Methodology 11

Combustion **Air Gasification**

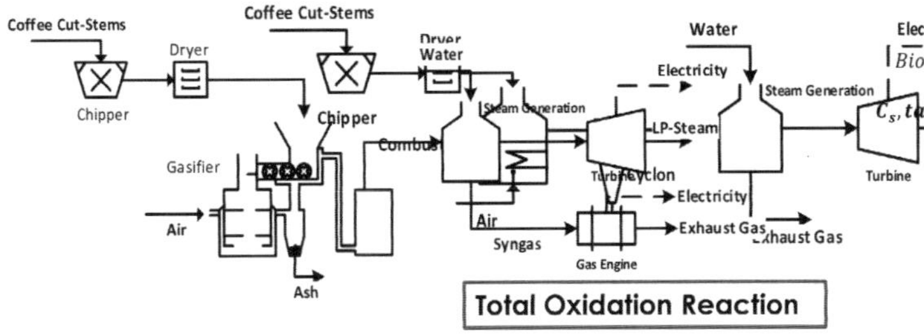

Pyrolysis

$Biomass + heat \rightarrow C_s + tar\ gases$

Combustion

$C_s, tar\ gases + O_2 \rightarrow CO_2 + H_2O + heat$

$tar\ gases + heat \rightarrow CO + H_2$

Reduction

$C_s + CO_2 \leftrightarrow 2CO$

$C_s + H_2O \leftrightarrow CO + H_2$

$C_s + 2H_2 \leftrightarrow CH_4$

$CH_4 + H_2O \leftrightarrow CO + 3H_2$

$CO + H_2O \leftrightarrow CO_2 + H_2$

Total Oxidation Reaction

$$Biomass + Air \rightarrow CO_2 + H_2O_{vapor} + N_2 + Heat$$

Reaction Mechanism

Physicochemical Characterization

3. Results 12

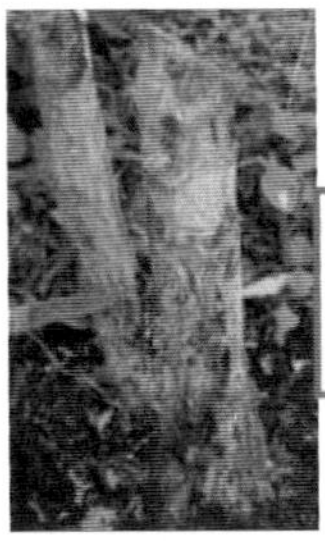

Raw Material	HHV (Mj/kg)
Sugarcane Bagasse	17.70
Rice Husk	16.47

	Coffee Cut-Stems
Moisture Content (%wt)	8.7
Chemical Composition (%wt dry)	
Cellulose	40.39
Hemicellulose	34.01
Lignin	10.13
Extractives	14.18
Ash	1.27
Proximate Analysis (%wt dry)	
Volatile Matter	82.15
Fixed Carbon	16.78
Ash	1.07
Elemental Analysis (%wt dry)	
Carbon	48.35
Hydrogen	5.93
~~Oxygen~~	~~44.21~~
HHV (MJ/Kg)	19.32

Gas Composition

3. Results 13

Outlet Gas Composition

Composition (Mass Fraction)	Combustion	Gasification
Steam	10.7	7.3
Oxygen	1.6	9.53E-6 ppm
Nitrogen	63.7	46.9
Carbon Monoxide	0.0008 ppm	24.9
Carbon Dioxide	23.9	19.1
NOx	18.43 ppm	3.85E-5 ppm
Hydrogen	------	1.8
Methane	------	1.4

Ministry of Environment, Housing and Territorial Development

Resolution 909 of 2008

Nitrogen Oxides Emissions of 0.0425 ppm

Energy Analysis

3. Results 14

Products	Units	Combustion	Gasification
LP Steam (3 bar)	Kg/h	840	455
Power	kW	61,7	647.5

→ Synthesis Gas

↓

Platform for chemicals and bioenergy production

Environmental Assessment

3. Results 15

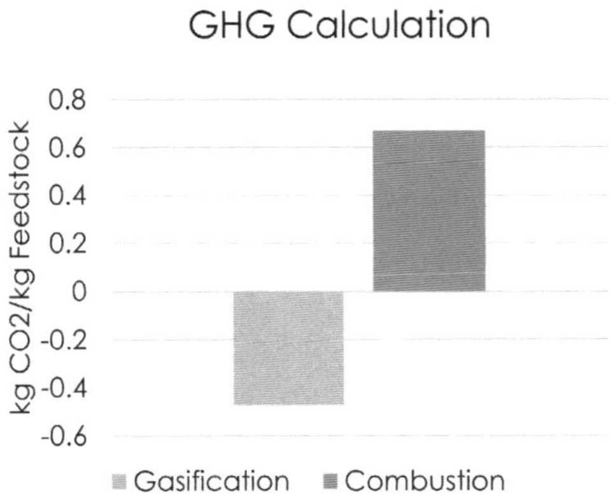

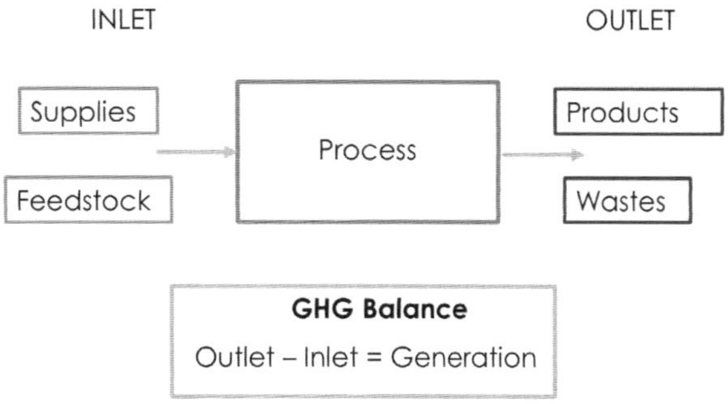

Conclusions

16

- Colombia is a country with a great diversity of residues from its long agricultural tradition that makes it a country with huge potential for converting these residues into high value-added products

- Biomass gasification seems to be the most promising technology for the use of Coffee Cut-Stems with high energy yields and low environmental issues. However, the combustion, for being a process with a low capital investment, has more applicability to industrial level than gasification technology.

Acknowledgement

17

- The authors of this work acknowledge to the National University of Colombia at Manizales, the Research Coffee Center (Cenicafé), the announcement 703 "Joven Investigador" of COLCIENCIAS and the Dirección de Investigación de Manizales (DIMA) for the financial support to developed this work.

References

18

- M. J. Valencia, C. A. Cardona, and C. Younes, *Cálculo de gases de efecto invernadero en procesos químicos y biotecnológicos con énfasis en cadenas de suministro*, Universidad Nacional. Manizales, Colombia, 2013.
- National Coffee Research Center, "Cenicafé Generates Electricity from Coffee Tree Wood," 2015.
- H. Balat and E. Kirtay, "Hydrogen from biomass - Present scenario and future prospects," *Int. J. Hydrogen Energy*, vol. 35, no. 14, pp. 7416–7426, 2010.
- J. A. Ruiz, M. C. Juárez, M. P. Morales, P. Muñoz, and M. a. Mendívil, "Biomass gasification for electricity generation: Review of current technology barriers," *Renew. Sustain. Energy Rev.*, vol. 18, pp. 174–183, 2013.
- H. C. H. Yang, "Gasification for Synthetic Fuel Production. Fundamental Process and applications," *Process Appl.*, 2014.
- M. Asadullah, "Barriers of commercial power generation using biomass gasification gas: A review," *Renew. Sustain. Energy Rev.*, vol. 29, pp. 201–215, 2014.

19

Comparison of the synthesis gas production via gasification and combustion from Coffee Cut-Stems

15AIChE
Annual Meeting, Salt Lake City, UT

Carlos A. García, Carlos O. Tascon, Carlos A. Cardona
e-mail: caagarciave@unal.edu.co
Universidad Nacional de Colombia. Sede Manizales

432007 Development of Kinetic Models of Biomass Gasification

Abstract:

This investigation provides a comprehensive analysis of simplified kinetic model for biomass gasification, over a wide temperature range (1150-1350°) in CO_2, H_2O and the combination of these two reactant gases. And the model simulates pyrolysis and gasification stages for a single biomass particle of three different types (poplar wood, corn stover and swichgrass) and different sizes. With the power-law model and non-linear least squares regression method, the kinetic model is optimized and the regressed kinetic parameters (pre-exponential factor and activation energy) are derived for both H_2O and CO_2. A = 0.0025 (g/m^2 s Pa) and E/R = 5179 K are the optimal parameters we obtain from these data, in order to predict the mass loss and the reaction rate. Model simulations follow great agreement with measured data.

Some publications investigated the influence of particle shape and size, including three main shapes, sphere, cylinderl and flat plate. Previous works reported that spherical particles react most rapidly, and flat particle react most slowly. The high ratio of surface-area-to-volume explains this observation. In order to better illustrate the influence of particle shape and size on the reaction rates, the spherical-equivalent-diameter is introduced in this investigation. This parameter is the diameter of a sphere derived with the same volume of any aspherical particles. The data shows that the normalized spherical-equivalent-diameter (d/d0) starts from 1 at the beginning of the reaction, and decreases to about 0.65 at the end of the reaction. Similar results are obtained for the other two biomass materials.

Ash effect is another important factor in the model. Based on the experimental data, a negative effect of the ash has been observed on the reaction rate. Some publications suggest that ash act as catalyst in gasification reactions, as a lot of researchers have observed this phenomena in their work. However, the experimental results indicate that the high-ash fuels react more slowly rather than more rapidly than the low ash fuels – the opposite of a catalytic effect. In conclusion, not all ash compositions are catalysts in the biomass gasification reactions. Even if the ash components have catalytic effects on the reactions, the effects of ash excluding available surface area has overwhelmed the catalytic effects on the reaction rates, as the reactions are kinetic limited in the experimental conditions. Thus, the net effect is a significant decrease rather than an increase in reactivity of biomass char, indicating that those catalytic effects could only be included by a compensating change in the ash exclusion model, if ash content is acting as catalyst in biomass gasification reaction.

More sophisticated models that dealing with surface adsorption, such as Ely-Reidel or Langmuir-Hinshelwood are compared in this investigation. From the observation, the simulated model fits the measured data without using the other terms in those complex expressions, which indicates that those terms are negligible compared to the dominant factors.

The uniform regressed parameters used in optimizing the global kinetic model works for most data in great correlation. And the regressed kinetic parameters explains why biomass gasification reaction proceeds more rapidly in H_2O environment than in CO_2 environment under otherwise similar conditions.

The conclusions from the experimental data are as follow:

1. Biomass raw materials in our experiments are mainly obtained from heartwood, corn stover and switchgrass, with an extensive range of both organic and inorganic components, nevertheless, no cardinal differences appear in the gasification parameters of char.
2. Under the same reacting conditions except reactant gas components, char gasification indicates different reaction rate, without any exception, reactions proceeds more rapidly with H_2O than with CO_2.
3. Over a broad range of temperature (1150°C-1350°C), and for different concentrations of CO_2, H_2O, and for the combination of CO_2 and H_2O, the first-order global model discussed in this study is accurate enough to optimize the kinetic rate data, which means the assumptions excellently satisfy the biomass gasification mechanisms.
4. Some advanced models need to be developed for high concentration of H_2O, however, that will increase the cost of optimization, and even none of the terms in those expressions can be precisely optimized. Thus, those added terms play minor roles in this model.
5. Some publications suggest that CO and H_2 are absorbed on particle surface. Howbeit, no statistically significant absorption is indicated, which affects the gasification reaction rates. Therefore, the diffusion of CO and H_2 molecules are negligible in this model.
6. The ash exclusion model indicates that although positive catalytic effects are reported in some publications, the negative effects has overwhelmed the catalytic effects. And the net effect is a significant decrease rather than an increase in reactivity of biomass char The pre-fixed experimental conditions determine that the reaction is kinetically limited, not diffusion limited, indicating that advanced and accurate surface-reaction-based model is not necessary.
7. The changing of biomass particle shape and size influences the reaction rates in profound ways, which cannot be described with any simple model trend. Better assumption and treatment should be developed to fit the data in this stage.

430912 Process Model of Biomass Gasification Using Aspen Simulation

Abstract:

To meet the demand of increasing energy needs, our current focus is on developing a commercial biomass gasification process. Efforts to scale the smaller scale laboratory reactor to commercial operation relies on first developing a comprehensive Aspen based process model, identifying the key operating parameters for the reactor and then optimizing the overall process. The process model attempts to mimic the actual biomass gasification system at steady state by analyzing each of the four zones inside a downdraft gasifier including: 1) Drying, 2) Pyrolysis, 3) Combustion and 4) Gasification zone. Using this multi-zonal approach, the process model can be easily modified for different operating facilities and conditions.

The current model is used to analyze the following important operational aspects of the gasifier system:

a) Syngas produced.

b) Liquid tar produced and gaseous tar present in the syngas.

c) Overall Energy Balance.

d) Air to Fuel Ratio.

e) Char Conversion.

f) Temperature profile through the gasifier.

Primary interest has been paid to the energy balance. The system, at steady state must be capable of operating without any external heat added to maintain the desired gasifier temperature profile. Energy release in the combustion zone must provide heat duty to support the other three processes occurring in the system. For this reason, reactors describing each processes inside the gasifier are kinetically modeled using a CSTR with specific surface and volumetric reactions. The solid-gas reaction rates are described using a shrinking core model.

ASPEN process parameters were identified to match different operating factors and used to optimize the overall process. Results were verified using experimental yield data collected from a lab scale biomass gasifier operated at the Missouri S&T Hybrid Energy Center and compared to previously published gasifier data.

435359 Multi-Phase CFD Simulation of a Downdraft Biomass Gasifier

Abstract:

A reliable, affordable clean energy supply is of primary importance for the economy and the environment and will prove to be pivotal in the 21st century. Biomass energy which is now accepted as a propitious alternative includes a widely available, sustainable and as a neutral CO2 resource, suited for modern applications for electricity generation, fuels and chemicals.

A 2D and 3D Mathematical model is developed to understand the configuration in a fixed bed biomass gasifier considering drying, pyrolysis, combustion and gasification reactions. The CFD simulation is based on the down draft biomass gasifier design run by Energy Research and Development center at Missouri University and Science and Technology, Rolla. A multi-phase Computational Fluid Dynamics model will be developed and verify against experimental setup. The flow model is based on an Eulerian –Eulerian description of the phases where the kinetic theory of granular flow forms the basis for the turbulence modelling in the solid phases and standard κ-ε turbulence model for gas phase.

Influence of the air ratio on gasification performance, such as production of syn gas, amount of char produced, and cold gas efficiency is presented with regard to the gasifier. The gas temperature distribution and product gas composition will also be considered into account. A comparison between the simulation and experimental data will be shown to accurately identify the simulation, confirming the validity of the current model.

435165 Comprehensive Study of Biomass Pyrolysis and Gasification Kinetics, Pore Stucture and Surface Area Modeling

Abstract:

Biomass gasification process is comprised of four stages, drying, devolatilization or pyrolysis, oxidation and gasification. Pyrolysis is a thermochemical decomposition of organic material at elevated temperatures in the absence of oxygen (or any halogen). It involves the simultaneous change of chemical composition and physical phase. Pyrolysis is a set of complex reactions observed in organic materials exposed to high temperatures. The pyrolysis stage is a really rapid process, usually less than 1 minute. And GC-MS (Gas chromatography–mass spectrometry) is set up to collect continuous data for the production gases during pyrolysis stage.

The experimental data in this investigation illustrates that the residence time of pyrolysis process is around 40 seconds. Another important result from this model is the fraction of mass loss at the end of pyrolysis. Typically, the fraction of mass loss for poplar is the highest among the three types of biomass material, usually between 92% and 96%. And this fraction is about 88%-90% for switchgrass and corn stover.

The pyrolysis process involves a huge amount of complex reactions and product gases. In the experiments, CH_4, C_2H_2, CO, NO, CO_2, benzene and toluene are on the list of my interests. CO is a little tricky when analyzing the GC-MS data, because N_2 and CO_2 has the same molecular weight and mass-to-charge ratio (m/z). To overcome this difficulty, argon is acted as purge gas to create an atmosphere free of N_2. A typical data for the pyrolysis gases analysis indicates that H_2O appears first, followed rapidly by benzene and toluene, then by C_2H_2, CH_4 and NO.

Char reactivity is demonstrated to be correlated with the char porous structure and surface area. The rate determining step in biomass gasification is the char-CO_2 or Char-H_2O reaction. In this report, the char conversion is introduced, to determine the reactivity of biomass chars. Char conversion is the ratio of the instant mass loss value and the mass loss value at the end of devolatilization. As biomass chars of different type and different size will have different total reaction time and reaction rate, it's more obvious to compare the reaction rates at a same char conversion, instead of residence time or other parameters. Internal surface area data are collected by N_2 adsorption, for chars produced at different reacting conditions (different concentrations of CO_2 and H_2O and different temperatures). Models have been developed to describe the surface area and pore distribution of biomass chars, which also illustrate the intrinsic rate of biomass gasification.

Some publications suggest that CO_2 adsorption gives different results with N_2 adsorption. Thus, CO_2 adsorption tests have been performed to the same biomass chars, and the surface area are applied to develop this model.

434169 ASPEN Plus Simulation of Combined Biomass Gasification and Fisher-Tropsch Synthesis of Liquid Fuels

Abstract:

APSEN Plus Simulation of Combined Biomass Gasification and Fisher-Tropsch synthesis of liquid fuels

Mohammad Rafati, Lijun Wang, Abolghasem Shahbazi

Abstract

Combined biomass gasification and subsequent Fischer-Tropsch (FT) synthesis is a promising pathway to replace the fossil-based fuels. In a FT process, a mixture of H_2 and CO (syngas) is converted to dominantly olefins and paraffins with various carbon chain-length in the presence of typically a Fe- or Co-based catalyst. CO_2 is a major constituent of syngas derived from biomass gasification which it typically removed from syngas with expensive and energy intensive separation unit operations prior sending the syngas to the FT reactor. CO_2 could be reactive in presence of Fe-based catalysts as Fe exhibits water-gas shift activity, in contrast to Co-based catalysts. We have developed and examined several process flow diagrams of combined gasification and FT synthesis for the production of liquid fuels from biomass using the ASPEN PLUS process modeling package. Intrinsic kinetic rate equations and detailed product distribution models of Fe- and Co-based catalysts are implemented into a FT reactor block using FORTRAN subroutines to accurately compare the effect of FT catalysts on process product distribution. The effects of various process parameters including catalyst type and recycling ratio on carbon conversion efficiency, energy conversion efficiency, and CO_2 emissions are examined.

409580 The Study on the Kinetics of Coal-Steam Catalytic Gasification for Different Particle Sizes of Coal and Catalyst

Abstract:

Compared with traditional gasification technologies, catalytic gasification technology can increase gasification rate obviously and reduce gasification temperature. So some researchers have paid more attentions to the catalytic gasification technology. The effects of particle size on the steam gasification of the coal char were investigated by the bench equipment. The gasification reaction of coal char and catalyst (K2CO3) with different diameter ranges were calculated with dynamic model to obtain the reaction rate constant for coal-steam gasification characteristics. Firstly, the non-catalytic gasification results of SF char with different particle sizes were analyzed under the condition 650 °C . It's found that particle size does not obviously affect the gasification reaction rate. Secondly, for the non-catalytic gasification results of SF char with same particle size under 650 °C and different gas rates, it was found that external diffusion can be ignored when the gas rate larger than 100ml/min. Then, catalytic gasification experiments of different chars, SF, YN, ZY, and the potassium carbonate with three particle sizes were conducted under 650 °C. The three experiment conditions are as follows. K2CO3 particle size are 75um, and the char particle sizes change; the char particle size is 200 um, and K2CO3 particle size changes; the char and K2CO3 particle sizes change at same range. The experimental results showed that the smaller the particle size of coal char and catalyst, the higher the gasification reaction rate under the experiment condition. For the catalytic gasification process, the influence of the catalyst particle size on catalytic gasification reaction is better than that of coal char particle size. When the granularity reaches more than 200 mesh, particle size of coal tar and catalyst change in the same range will have a best effect on catalytic gasification, and it is better when the catalyst particle size changes than that of coal char particle size changes also. At last, the mobility of K2CO3 in gasification reaction and the influence of catalyst particle size on gasification reaction are discussed.

432057 A Novel Enzyme-Catalyzed Process for Biodiesel Production

Abstract:

Lipase-catalyzed transesterification (also esterification) from oils or fats for biodiesel production has some advantages over chemical-catalyzed transesterification, such as environmental friendship, lower energy consumption being widely suitable to low quality oils or fats. However, the low stability of the lipase due to alcohol (methanol) has been being regarded as the main hurdle to the industrialization of lipase-catalyzed biodiesel production. Tsinghua University has developed a novel process to eliminate the inhibition of methanol on lipase, thus the operational life of the lipase could be improved into more than 300 cycles from several (1-5) cycles while traditional process applied. The patented novel process was successfully demonstrated in a pilot plant (200kg/day) and commercial plant (50,000ton/year) in China. It has also been being demonstrated in a pilot-plan in Brazil.

During the past 20 years, it has been being an attractive project to research and improve the enzymatic process for biodiesel production. Of course, the enzyme (lipase) itself is one of the core factors. A lot of research focused on how to improve the specific activity or stability of lipase. Its stability depends upon not only enzyme itself but also how to use it. Some researchers use the free lipase (which is cheaper) as catalyst, but do not recycle lipase, other researcheres use and recycle immobilized lipase which is much expensive. This work presents a novel process in which both free and immobilized lipase are applied and successfully recycled. The free lipase will first catalyze more than 90% of the conversion of oils and the obtained products contain still free fatty acid (FFA) which is not accepted by the standard specification of biodiesel. Then the immobilized lipase will further catalyze the conversion of the unconverted feedstock form the first step.

Compared to the process using only free lipase (normally a neutralization with alkaline needed to remove FFA), the novel process can directly get the qualified products and higher yield because of the conversion of FFA. Compared to the process using only immobilized lipase, the dosage of immobilized lipase can be significantly reduced with the novel process because that less than 10% of the conversion is left for the immobilized enzyme after the first conversion with free lipase.

434779 Co-Liquefaction of Lignocellulosic Biomass and Heavy Oil Residue By One-Step Reaction Using Supercritical Organic Solvents

Abstract:

Co-processing of biomass and biofuels with conventional petroleum feedstock using already installed oil refinery processes seems to be a promising alternative for bio-oil production. In this study, simultaneous one-step co-liquefaction of wood biomass (microcrystalline cellulose or oil palm empty fruit bunch fibers (EFB)) and vacuum residue (VR) was carried out by using sub- and supercritical solvents such as m-xylene, n-dodecane, tetralin and decalin.

The co-liquefaction performance with/without the Fe_3O_4 catalyst and initial H_2 pressure in each solvent was evaluated with conversion, liquid yield, coke formation and product distribution, and was compared with the results of only VR liquefaction. A simulated distillation (SIMDIS) gas chromatography method was used for the analysis of liquid products. In the absence of catalyst and H_2 at 450°C, the co-liquefaction of VR and biomass could reach up to 88.5% while the coke amount did not exceed 5%. The Fe_3O_4 catalyst significantly contributed to suppressing coke formation and thus increased liquid product yield in all solvents (especially in m-xylene and n-dodecane) if H_2 was also introduced to the reaction. The conversion in co-liquefaction could be improved only by VR upgrading reaction, and EFB was more favorable than microcrystalline cellulose in liquid yield and coke suppression. The obtained liquid products were completely soluble in low polar solvents.

421473 Advanced Biofuel from Various Lignocellulosic Feedstocks By Liquid Phase Pyrolysis - the Biocrack Pilot Plant

Abstract:

Advanced biofuel from various lignocellulosic feedstocks by liquid phase pyrolysis - the bioCRACK pilot plant

[1]J. Ritzberger, [1]P. Pucher, [2]N. Schwaiger, [2]M. Siebenhofer

[1]BDI – Bioenergy International AG, Parkring 18, 8042 Grambach, Austria, e-mail: juergen.ritzberger@bdi-bioenergy.com

[2]Graz University of Technology, Institute of Chemical Engineering and Environmental Technology, Inffeldgasse 25/C/II, 8010 Graz

Currently almost all commercial available biofuels compete with food and animal feed production. Advanced biofuels are expected to be produced from non-food feedstock. They will play an important role in generation of renewable liquid energy carriers. The bioCRACK process represents a biomass-to-liquid concept for generating advanced biofuels, technologically based on liquid phase pyrolysis. Spruce wood, wheat straw, beech wood and miscanthus is the selected lignocellulosic feedstock. Preferred heat carrier is vacuum gas oil, an intermediate from vacuum distillation of crude oil. Lignocellulosic biomass is processed in vacuum gas oil at temperatures of 350°C to 400°C and ambient pressure. During liquid phase pyrolysis the feed is transferred into hydrocarbons, liquid CHO-products, reaction water, biochar and gaseous products. Technology as well as the type of heat carrier suggest integration of the process in an oil refining concept. Crude products from liquid phase pyrolysis and the spent heat carrier can be upgraded in existing facilities of the refinery. For instance, the crude product is upgradeable to diesel fuel via hydrogenation.

From 2012 to 2014 a fully integrated bioCRACK pilot plant with a design capacity of 100 kg/h was in continuous operation at the OMV oil refinery in Schwechat/Austria. Within two and a half year of intensive research the influence of various lignocellulosic feedstocks (spruce wood, beech wood, wheat straw and miscanthus), different biomass impregnation (high and low boiling refinery intermediates), operation parameters, yield and composition of the products, the mutual interaction of the feed and the heat carrier and the impact of process conditions and products on the refining process have been investigated.

Within the design temperature span (350 °C to 400 °C) the liquid phase pyrolysis of the deployed feedstocks showed comparable results. Mass balances and C^{14}-tracking confirmed bio-carbon transfer from lignocellulosic feed into crude product fuels of 10 to 20 %. 9% to 18 % of the bio-carbon is transferred into the heat carrier. It was shown that the combined conversion of lignocellulosic and vacuum gas oil by liquid phase pyrolysis has an accelerating impact on the decomposition of the heat carrier oil. In conclusion the bioCRACK process may significantly contribute to providing advanced biofuels from lignocellulosic biomass as well as generation of low molecular weight intermediates from vacuum gas oil.

433704 Catalytic Ethanolysis of Kraft Lignin for the Production of Aromatic Chemicals over Zeolite

Abstract:

Due to dwindling fossil feedstock, lignocellulosic biomass is becoming an important alternative, renewable resource for the production of fuels, chemicals, and carbon based materials. Lignin has been a high-volume byproduct of the paper and pulp industry since the beginning of the commercial operation of alkaline pulp technology. Here, we demonstrate ethanolysis of lignin in supercritical ethanol (280 ^{o}C) over zeolite. With zeolite Y as catalysts, the high-value chemicals aromatics are accounted for around 50% of the products. And the mostly aromatics are toluene and propyl benzene.

(675e) Hydrodeoxygenation of Oxidized Low-Water Bio-Oil Produces Higher Hydrocarbon Yield

Luo, Y.,
Guda, V. K., Mississippi State Univeristy
Steele, P., Mississippi State University
Originally presented on: 11/12/2015 13:54:00 - 14:15:00

Fast pyrolysis bio-oil has a high oxygen content of 40-45% in the form of oxygenates and approximately 30 vol. % water content; both lead to numerous negative properties. Bio-oil needs to be upgraded for use as liquid transportation fuel. Hydrodeoxygenation is one of the most effective methods to remove oxygen. However, catalysts usually employed have been rapidly deactivated because of high water content and coking resulting from charring of the highly reactive oxygenates. In this study, we tested lower water content bio-oil for hydrodeoxygenation as means to reduce coking and enhance hydrocarbon yields. Distilled bio-oil and oxidized distilled bio-oil were hytrotreated and hydrocracked with nickel/silica-alumina catalysts. Raw bio-oil and oxidized raw bio-oil had the same hydroprocessing treatments applied. All four hydrocracked products contained a water content of nearly zero, a higher heating value of 44-45 MJ/Kg and an oxygen content near zero.The oxidized bio-oils had higher hydrocarbon yields than did non-oxidized bio-oils. Gas chromatography/mass spectrometry characterization showed that 58-64% of aromatic hydrocarbons and 36-42% of alkyl hydrocarbons were the main products for both distilled bio-oil and raw bio-oil while 63-70% of alkyl hydrocarbons and 30-37% of aromatic hydrocarbons were the main products for both oxidized bio-oils.

Using Pulsed Electric Field Pre-treatment to Improve Methane Production

Seyedeh Masoumeh Safavi and Runar Unnthorsson,
School of Engineering and Natural Sciences, University of Iceland, Iceland

Abstract

The efficiency of the anaerobic digestion (AD) process mostly depends on the bioavailability of the organic solids. Through specific pre-treatments of the substrate the AD process will convert the organic solids better and faster. Laboratory-scale batch AD tests were run to analyze the influence of pulsed electric fields on the methane production from pig slurry (PS), landfill leachate (LL) and fruit/vegetable waste (FVW). The results were compared with untreated samples (control). Methane production of the PS and LL compared to the control was significantly increased up to 50 and 48%, respectively, with the highest intensity (180 kJ/kg). The result of PEF pre-treatment on FVW showed that lowest intensity (54 kJ/kg) was the intensity by which the highest amount of methane (up to 13%) was achieved. Chemical oxygen demand removals were increased up to 45, 100 and 17% for PS, LL and FVW, respectively, compared to the control.

Introduction

The produced amount of organic waste (consisting of municipal, industrial, and agricultural waste) is increasing dramatically each year which can lead to a significant environmental pollution. Considerable amounts of greenhouse gasses (GHGs) such as CH_4, N_2O and CO_2 are emitted during disposal and storage of organic waste contributing to global climate change (Wolter, Prayitno, and Schuchardt 2004). Anaerobic digestion (AD), a well-known waste treatment technology, converts organic waste into the useful and valuable product (methane) that can be used as a substitute for fossil fuels, and help societies towards transportation fuel independence. It can also be used to generate electricity and heat (Krishania et al. 2013). Furthermore, the EU policy concerning renewable energy sources has a target of supplying 20% of the European energy demands from renewable energy, and also prohibition of landfill using organic matter by 2020. The question arises here whether it is possible to improve the performance of AD and reach a higher yield of methane production?

In order to partly answer this question, authors investigated, in this paper, the effects of pulsed electric field (PEF) pre-treatments of different substrates on the AD process. The hypothesis is that PEF pre-treatment improves the performance of AD contributing to higher methane production (Choi, Jeong, and Chung 2006; Lee and Rittmann 2011; Salerno et al. 2009).

The experimental work in this study was performed on three different substrates: pig slurry (PS), landfill leachate (LL), and fruit/vegetable waste (FVW). Analyses of the obtained results are based on comparison between treated and untreated substrates to study the effect of PEF pre-treatment on methane production. The studied PEF pre-treatment in this work uses different intensities of 54, 108 and 180 kJ/kg. The anaerobic digestibility of the pre-treated batches compared against control batches.

Materials and methods

Substrates and inoculum

The PS samples were collected from a pig farm located in Grundarhverfi, Reykjavik, Iceland. LL and FVW were collected from the municipal solid waste landfill in Alfsnes, Reykjavik, Iceland. FVW was shredded and blended with additional tap water. Prior to the PEF, PS and FVW were filtered through an 8 mm sieve to remove large pieces. The samples were stored below 5°C until used. The initial TS concentrations were 2.5, 0.70 and 3.9%, with a total VS content of about 70.7, 43.8 and 90.8% for PS, LL and FVW, respectively. COD values were adjusted to be in the range of about 11000, 11600 and 5000 mg/L for PS, LL and FVW, respectively. The inoculum was a mesophilic anaerobic sludge from the wastewater treatment plant of Vifilfell Company. Inoculum was degassed for 7 days before feeding it to the digester in order to eliminate the residual biodegradable organic content.

Substrate analysis and preparation

Samples were analyzed for total solids (TS), volatile solids (VS), chemical oxygen demand (COD), soluble COD (SCOD), total suspended solids (TSS) and volatile suspended solids (VSS) according to the Standard Methods [8].

Pulsed Electric Field pre-treatment

The PEF system was designed for treating the organic solids in the slurry. The equipment used is shown in Figure 2. Charging of the main circuit was done with a capacitor charging unit (HCK 200-20000, FuG Elektronik) which supplied one capacitor (40kV–1μF) with electrical energy. The field strength, the frequency and the pulse width were 20 kV/cm, 1.7Hz and 4μs, respectively. The slurry was pumped continuously from the feed tank, through the chamber and back into the feed tank. The electrical pulses were applied to the slurry in the chamber when it flowed past the electrodes that are positioned at each side of the chamber. The gap between the two electrodes was 1cm, and the breakdown voltage of the spark gap (18.5 KV). The partly treated slurry coming from the chamber was directed back to the feed tank. The slurry circulation continued until the required time for each treatment was reached. The required time was determined using the following equation:

$$T=V\times(TI/E) \tag{1}$$

where T (min) is the required time for the PEF treatment, V is the substrate volume to be treated (m^3), TI is the treatment intensity (kJ/kg) and E is the effective energy output of the equipment per hour (kWh/h).

Batch setup

BMP assays were performed on the treated and untreated samples with three different intensities, each in triplicate. The substrates were mixed with inoculum and water to get a ratio substrate (VS)/inoculum (VS) of 1. The quantities were calculated to get a final volume of 900 mL of liquid mixture. The test bottles were placed at 37 °C water bath. During the test period the biogas formed in the bottles was continuously conveyed by pressure difference to the cylinders with a pipe. All cylinders were immerged in the water and kept at the same pressure condition in order to ensure accurate measurement of gas volume. Produced biogas directed to the cylinders and raised them up. The distance between the top of the gas cylinders and the water level was used to measurethe volume of the produced biogas. The quality of the biogas was measured withthe GEOTECH GA5000 gas analyser.

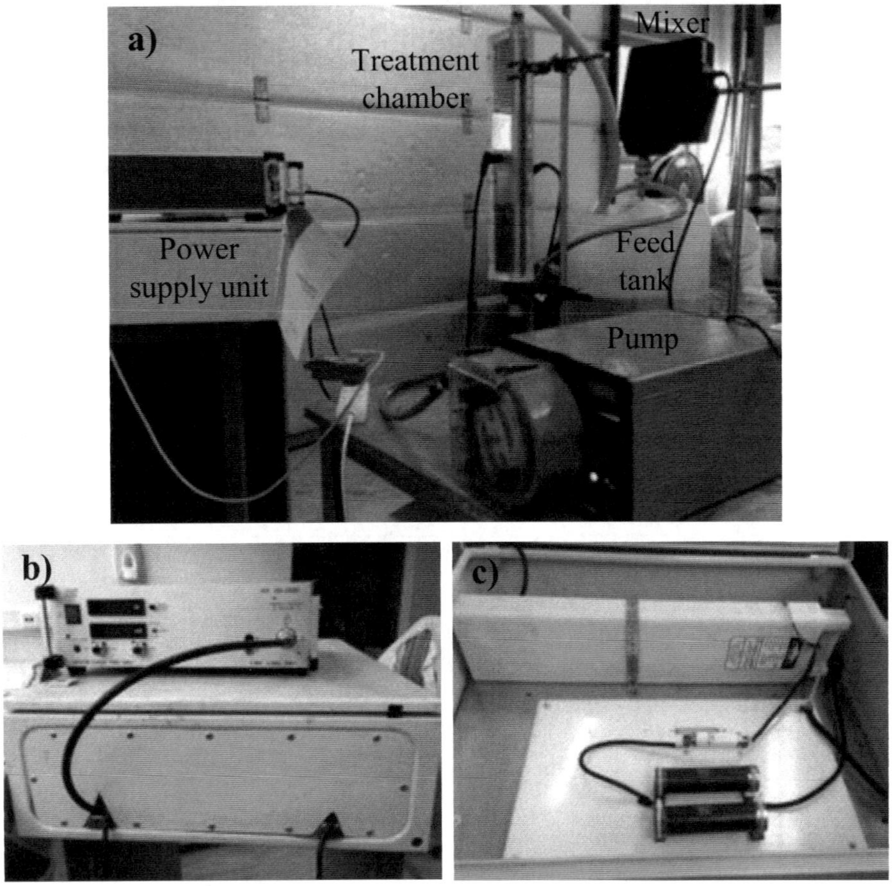

Figure 1: PEF instrument (a), energy supplier unit (b), capacitor, spark gap and resistors (c).

Result and discussions

Property changes by PEF pre-treatment

A summary of results from analytical measurements of the samples after PEF pre-treatment compared to the control samples are given in Table 1. The presence of insoluble organic solids in the samples was affected by the PEF pre-treatment. As can be seen in Table 1, PEF had positive effects on solubilization of FVW and PS, but it did not have any effect on LL with regard to the solubilization. The release of soluble components after the PEF pre-treatment was confirmed by decreasing the TSS content of the treated PS and FVW samples. The SCOD increased for all the conditions of FVW samples and only for condition 1 of PS samples.

Table 1: Effects of PEF pre-treatment on solubilization of organic solids compared to the control digesters. (0) no significant influence, (+) positive significant effect and (Na) not available. VSS analyses have been done for FVW samples, but because of the high volatility of the samples it could not be measured.

Substrate	PEF conditions	Characterization of samples		
		TSS	VSS	SCOD
Pig slurry	Condition 1	+	+	0
	Condition 2	+	+	+
	Condition 3	+	+	0
Landfill leachate	Condition 1	0	0	0
	Condition 2	0	0	0
	Condition 3	0	0	0
Fruit and vegetable waste	Condition 1	+	Na	+
	Condition 2	+	Na	+
	Condition 3	+	Na	+

Methane production

The enhancement percentage of methane production and COD removal of the PEF treated samples compared to the controls are depicted along with different levels of energy applied in Figures 2, 3 and 4 that are for PS, LL and FVW, respectively.

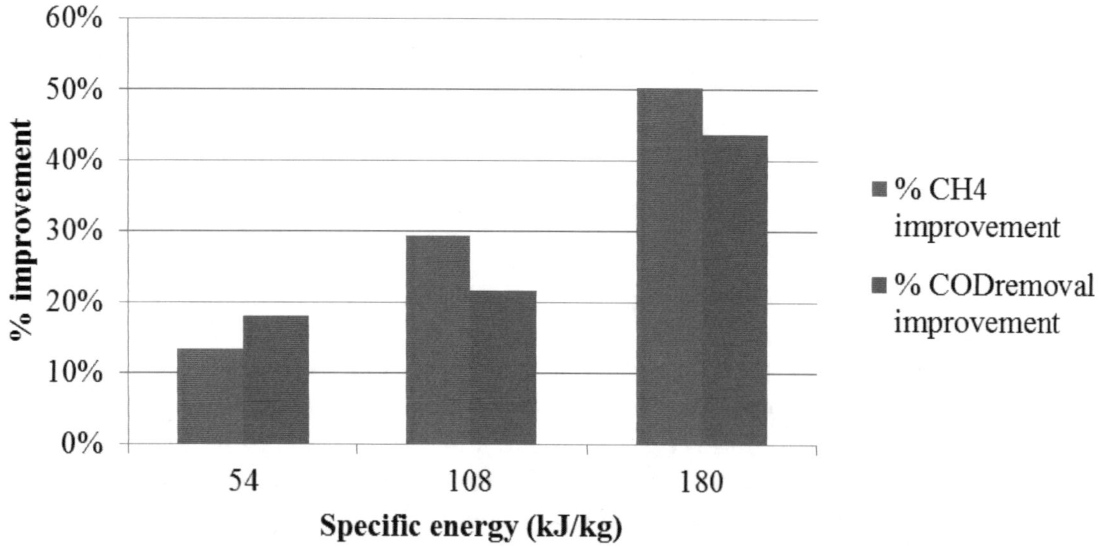

Figure 2: Improvement of the methane production and COD removal compared to the control digesters for pig slurry.

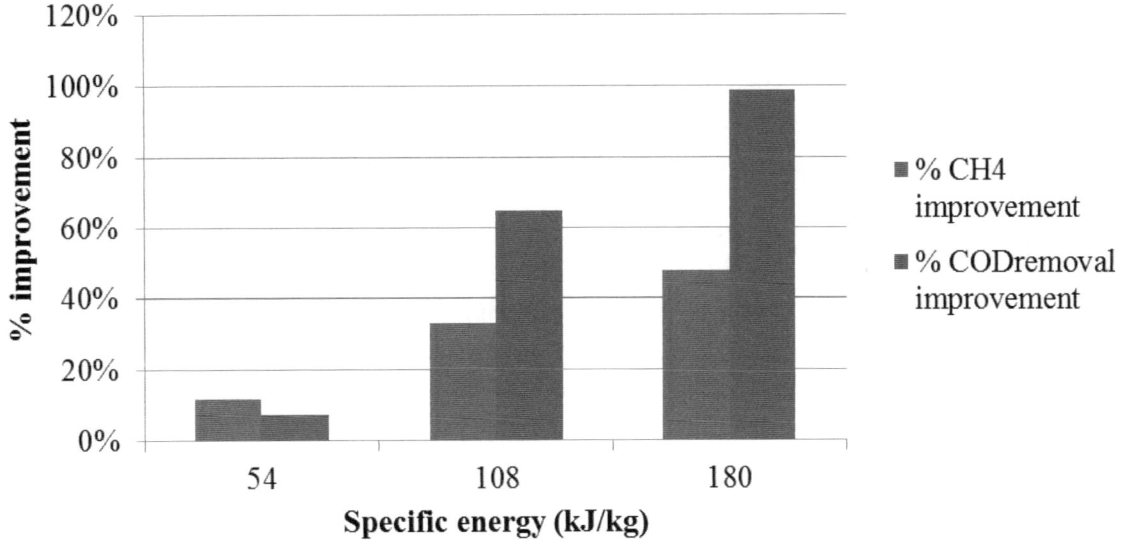

Figure 3: Improvement of the methane production and COD removal compared to the control digesters for landfill leachate.

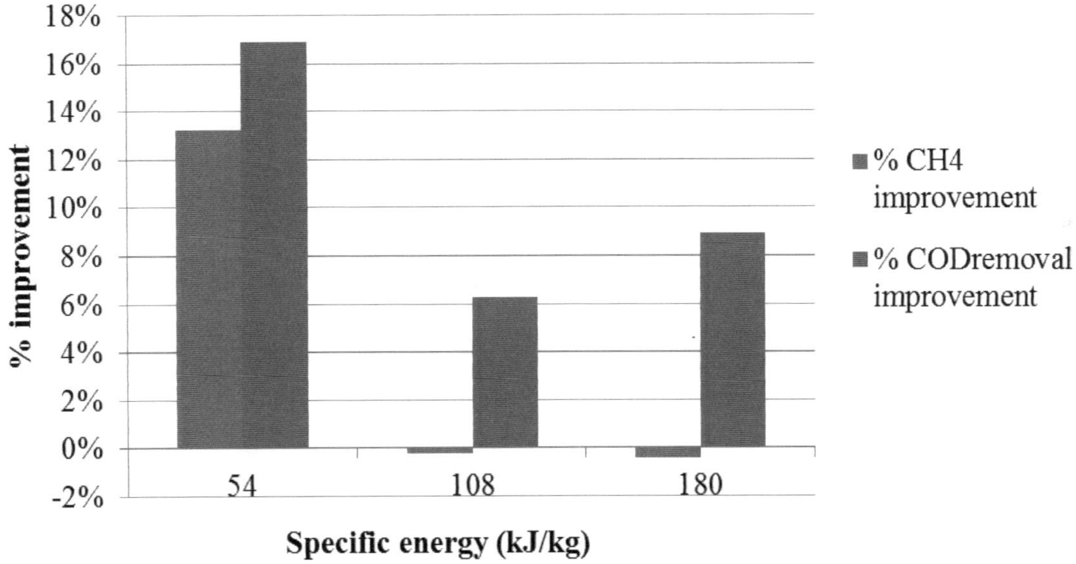

Figure 4: Improvement of the methane production and COD removal compared to the control digesters for fruit and vegetable waste.

Figures 2 and 3 show that the methane production and the COD removal did improve in every case for PS and LL compared to the controls. The results indicate the positive effect of PEF treatment on methane production potential. The methane yields were approximately 225 and 135 (ml/g COD) for the control PS and LL samples, respectively. The methane production after PEF pre-treatment represented up to 50% and 48% improvement compared to the controls for PS and LL, respectively. Thus higher PEF intensities could possibly increase the methane production from both PS and LL. The increase in methane production of the treated PS and LL samples in this study are higher than that acheived by Elbeshbishy et al. (Elbeshbishy et al. 2011) with ultrasonic pre-treatment of hog manure and Kopplow et al. (Kopplow, Barjenbruch, and Heinz 2004) with PEF pre-treatment of sewage sludge. Elbeshbishy (Elbeshbishy et al. 2011) reported only 28% increase in methane yield with intensity of 500 (kJ/kg) which is almost three times higher than the highest intensity in our experiment (180 kJ/kg). Kopplow (Kopplow et al. 2004) reached 20 and 5% enhancement in biogas yield with intensities of 150 and 280 (kJ/kg), respectively. Figure 4 summarizes our results for FVW. The initial methane producton from the control was approximately 1420 (ml/g COD). The results also showed that methane production increased by 13% after PEF treatment compared to the controls. Unlike PS and LL, the PEF treatment effect on methane production from FVW was more significant for the lowest intensity applied. PEF pre-treatment made solids more bioavailable, which led to COD removal of up to 100% for LL compared to the control. This is much higher than the 17% and 45% reduction obtained from FV and PS samples, respectively.

Conclusion

An experimental study with pulsed electric field (PEF) pre-treatment was conducted to enhance methane yield of pig slurry (PS), landfill leachate (LL) and fruit/vegetable waste (FVW). The influence of pre-treatment on methane production was tested via the batch anaerobic digestion (AD) test. The results of the PEF pre-treatment on all the substrates (PS, LL and FVW) showed positive effects on the methane production. The highest methane improvement was obtained from the PEF pretreated PS (with 50% increase) and the PEF pretreated LL (with 48% increase) , while only 13% improvement was acheived from the PEF pretreated FVW. Methane production of the PS and LL increased with the highest intensity (180 kJ/kg). On the contrary, FVW showed that lowest intensity (54 kJ/kg) was the intensity by which the highest amount of methane was achieved.

Acknowledgement

The authors would like to thank SORPA for the financial support and the Innovation Centre Iceland for the operation of the pulsed electric field equipment.

References

1. Choi, Hanna, Seung-Woo Jeong, and Youn-Jin Chung (2006), "Enhanced Anaerobic Gas Production of Waste Activated Sludge Pretreated by Pulse Power Technique," *Bioresource technology* 97(2):198–203.

2. Elbeshbishy, Elsayed, Saad Aldin, Hisham Hafez, George Nakhla, and Madhumita Ray (2011), "Impact of Ultrasonication of Hog Manure on Anaerobic Digestability," *Ultrasonics Sonochemistry* 18(1):164–71.

3. Kopplow, Ole, Matthias Barjenbruch, and Volker Heinz (2004), "Sludge Pre-Treatment with

Pulsed Electric Fields," *Water Science and Technology* 49(10):123–29.

4. Krishania, Meena, Virendra Kumar, Virendra Kumar Vijay, and Anushree Malik (2013), "Analysis of Different Techniques Used for Improvement of Biomethanation Process: A Review," *Fuel* 106:1–9.

5. Lee, Il-Su and Bruce E. Rittmann (2011), "Effect of Low Solids Retention Time and Focused Pulsed Pre-Treatment on Anaerobic Digestion of Waste Activated Sludge," *Bioresource technology* 102(3):2542–48.

6. Salerno, Michael B., Hyung-Sool Lee, Prathap Parameswaran, and Bruce E. Rittmann (2009), "Using a Pulsed Electric Field as a Pretreatment for Improved Biosolids Digestion and Methanogenesis," *Water Environment Research* 81(8):831–39.

7. Wolter, Martin, Shafiq Prayitno, and Frank Schuchardt (2004), "Greenhouse Gas Emission during Storage of Pig Manure on a Pilot Scale," *Bioresource Technology* 95:235–44.

434705 Economic Analysis of Simulated Biobased 1,3-Propanediol Production Process

Abstract:

Over-production of glycerol from biodiesel industry has led to a reduction in its sale prices. One way to make it profitable is carrying out a fermentative transformation towards a high value added product such as 1,3-propanediol. The production of this diol from industrial glycerol grade (80% w/w) was technically and economically evaluated using *Clostridium sp*. in fed-batch fermentation with an inorganic nitrogen source. Considering that fermentation broth is composed by several metabolites and biomass, solid-liquid phase was treated by centrifugation and liquid multicomponent phase separation was approached by evaporation and a distillation train stage. This latter was modeled using NRTL parameters regressed by data from literature. The process produces 1,3-propanediol with a purity greater than 99% by weight suitable for commercial use. A few variables were taken into account such as split of glycerol for batch and fed-batch stages, glycerol price, the total flow and concentration of glycerol fed in fermentation in order to evaluate economic process feasibility with an industrial scale approach and contextualized within the current state of market.

Mathematical and Experimental Modelling of Biomass Gasification for Hydrogen Production

Carlos A. García, Álvaro Gómez, Pisarenko Yu. A, Carlos A. Cardona
[a]Instituto de Biotecnología y Agroindustria, Departamento de Ingeniería Química. Universidad Nacional de Colombia sede Manizales
[b] Moscow State Academy of Fine Chemical Technology, Moscow –Russia.

Outline

- Introduction
  - Lignocellulosic Biomass in Colombia
  - Gasification
- Objective
- Methodology
  - Mathematical Modelling
  - Experimental Procedure
  - Simulation
- Results
- Conclusions
- References
- Acknowledgements

Pinus Patula

1. Introduction

Figure *Pinus Patula* Chips

- High Availability
- High Energy Potential
- Issues: Recollectation and transport logistics

Reforestation 17 Million Ha → Only 3% are being exploited → Reforested Area 510,000 Ha → Residues 13 Million tonnes

*Mining and Energy Planning Unit (UPME)

Coffee Cut-Stems

1. Introduction

Figure. Coffee Cut-Stems Chips

Colombia ist the fourth largest producer of Green Coffee

↓

9.9 million of bags (60 kg-bag) in 2013

↓

Residues from renovation 17 tonne per hectare

↓

Available 1.53 million tonnes/year

- Renovation: Every 4 – 5 years
- Competition with other technologies: Combustion
- Logistics and transport issues

Transformation Methods

1. Introduction

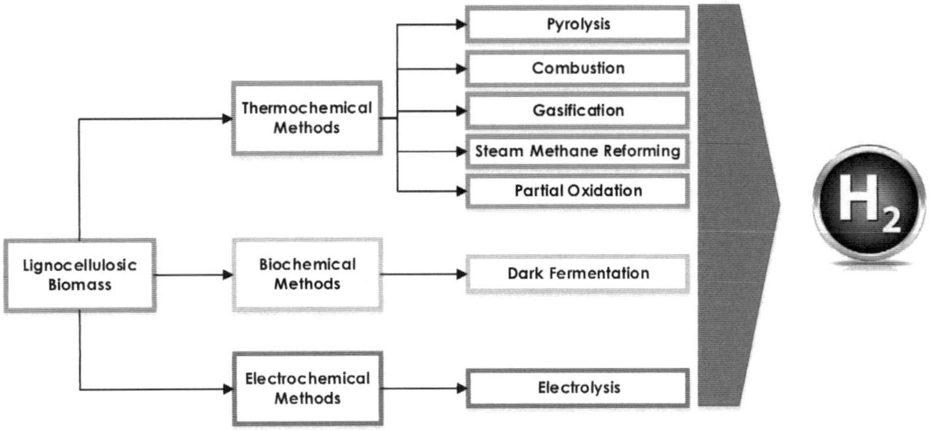

Hydrogen

1. Introduction

Promising Source of Energy

- High Energy Densitiy
- Low CO2 Emissions

Only 4% of Hydrogen is produced from Renewable Sources

Platforms for Hydrogen production

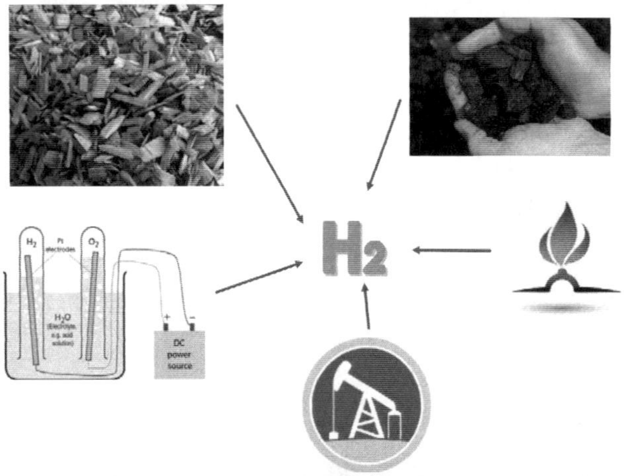

Mathematical Modelling

1. Introduction

Kinetic Models

N. Gao, A. Li, **Modeling and simulation of combined pyrolysis and reduction zone for a downdraft biomass gasifier**, Energy Conversion and Management, 2008, 3483–3490

$$r_1 = nA_1 \exp\left(\frac{-E_1}{RT}\right)\cdot\left(P_{CO_2} - \frac{P_{CO}^2}{K_1}\right)$$

$$r_2 = nA_2 \exp\left(\frac{-E_2}{RT}\right)\cdot\left(P_{H_2O} - \frac{P_{CO}\cdot P_{H_2}}{K_2}\right)$$

$$r_3 = nA_3 \exp\left(\frac{-E_3}{RT}\right)\cdot\left(P_{H_2}^2 - \frac{P_{CH_4}^2}{K_3}\right)$$

$$r_4 = nA_4 \exp\left(\frac{-E_4}{RT}\right)\cdot\left(P_{CH_4}\cdot P_{H_2O} - \frac{P_{CO}\cdot P_{H_2}^3}{K_2}\right)$$

$$r_5 = nA_5 \exp\left(\frac{-E_5}{RT}\right)\cdot\left(P_{CO_2}\cdot P_{H_2} - \frac{P_{CO}\cdot P_{H_2O}}{K_6}\right)$$

Thermodynamic Equilibrium Models

S. Jarungthammachote, A. Dutta, **Thermodynamic equilibrium model and second law analysis of a downdraft waste gasifier**, Energy, 2007, 1660–1669.

$$\begin{aligned} CH_xO_yN_z + wH_2O + m(O_2 + 3.76N_2) \\ = n_{H_2}H_2 + n_{CO}CO + n_{CO_2}CO_2 + n_{H_2O}H_2O \\ + n_{CH_4}CH_4 + \left(\frac{z}{2} + 3.76m\right)N_2, \end{aligned}$$

Neural Network Models

M. Puig-Arnavat, J. Hernandez, J. Bruno, A. Coronas, **Artificial neural network models for biomass gasification in fluidized bed gasifiers**, Biomass and Bioenergy, 2013, 279–289

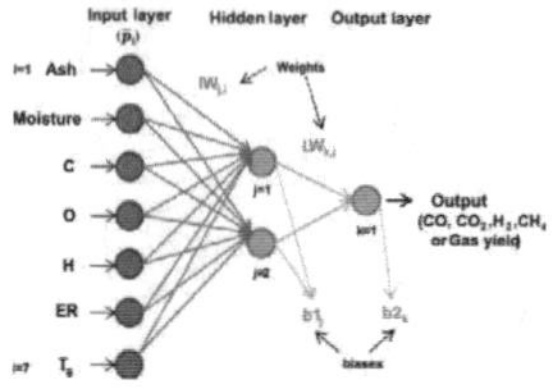

Gasification Behavior

- Moisture Content
- Air/Biomass Ratio

Objective

The aim of this work is to evaluate the effect of the moisture content and air/biomass ratio in the synthesis gas production as platform for the hydrogen production using two lignocellulosic residues in the Colombian case such as Coffee Cut-Stems and Pinus Patula in a downdraft gasifier.

General Description

2. Methodology

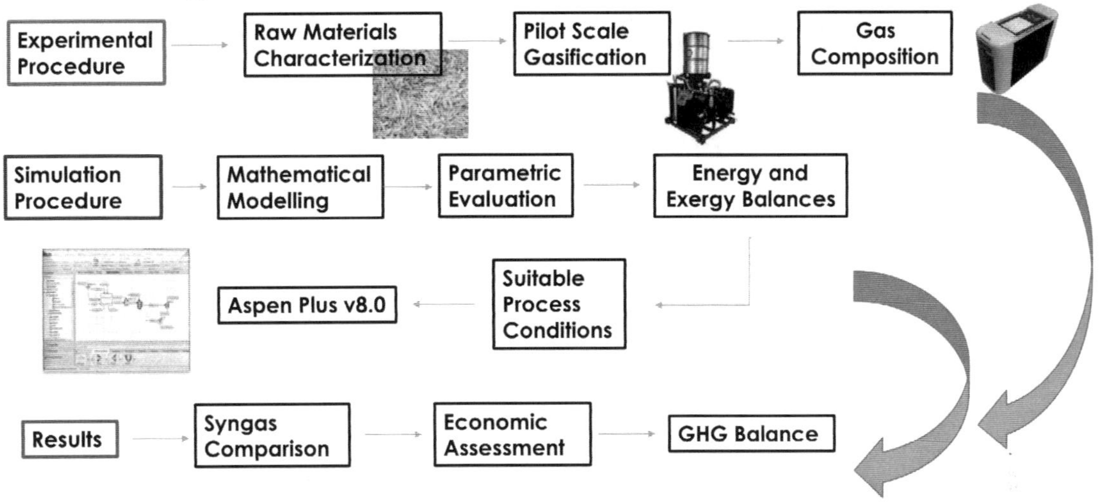

Raw Materials

2. Methodology

Chemical Composition

Elemental Analysis

- $C = 0.635 * FC + 0.460 * VM + 0.095 * ASH$
- $H = 0.059 * FC + 0.060 * VM + 0.010 * ASH$
- $O = 0.34 * FC + 0.469 * VM + 0.023 * ASH$

Calorific Value

- $HHV = 19.2880 - 0.2135\left(\frac{VM}{FC}\right) - 1.9584\left(\frac{ASH}{VM}\right) + 0.0234\left(\frac{FC}{ASH}\right)$

Pilot Scale Gasification

2. Methodology

Parameter	Value
Moisture Content	10 – 20%wt.
Particle Size	0.5 – 1 cm
Portable Gas Analyzer (%Vol.)	
Hydrogen	0 – 50%
Carbon Monoxide	0 – 40%
Carbon Dioxide	0 – 30%
Methane	0 – 10%
Oxygen	0 – 25%

10 kW Gasifier using air as gasifying agent

Chemical, Catalitic and Biotechnological Processes Research Group 11

Mathematical Modelling

2. Methodology

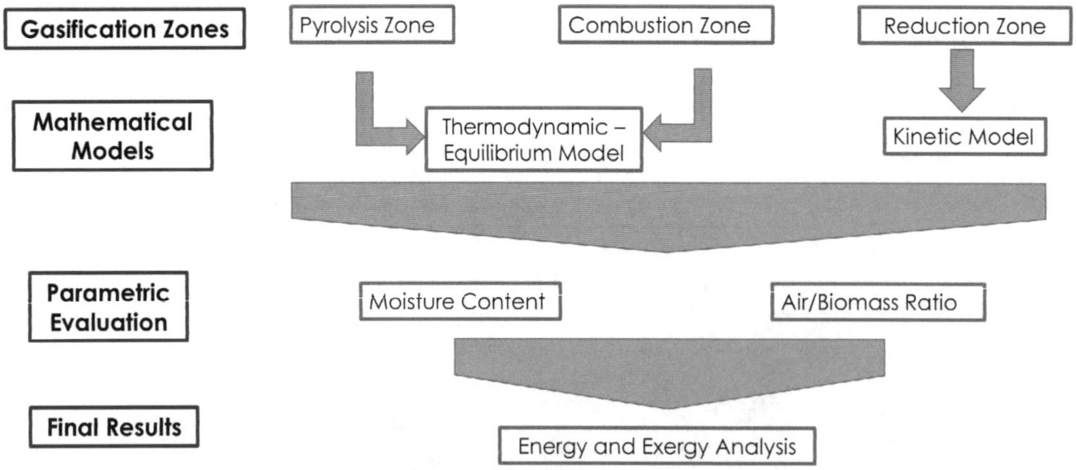

Chemical, Catalitic and Biotechnological Processes Research Group 12

Simulation

2. Methodology

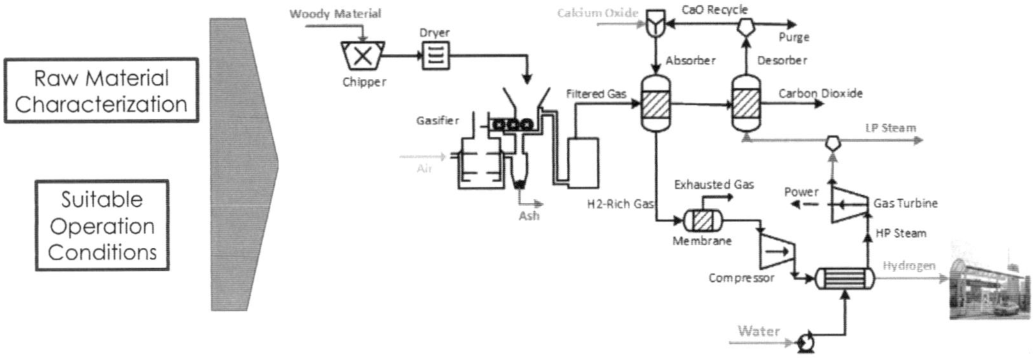

Physicochemical Characterization

3. Results

	Pinus Patula	Coffe Cut-Stems
Moisture Content (%wt)	9.21	8.7
Chemical Composition (%wt dry)		
Cellulose	44.78	40.39
Hemicellulose	23.75	34.01
Lignin	20.22	10.13
Extractives	11.0	14.18
Ash	0.25	1.27
Proximate Analysis (%wt dry)		
Volatile Matter	82.14	82.15
Fixed Carbon	17.64	16.78
Ash	0.23	1.07
Elemental Analysis (%wt dry)		
Carbon	48.96	48.35
Hydrogen	5.97	5.93
Oxygen	44.31	44.21
HHV (MJ/Kg)	19.97	19.32

Raw Material	HHV (MJ/kg)
Sugarcane Bagasse	17.70
Rice Husk	16.47

Effect of Moisture Content

3. Results

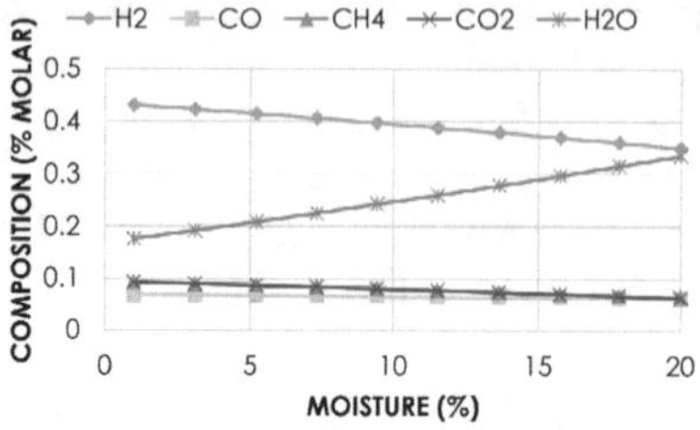

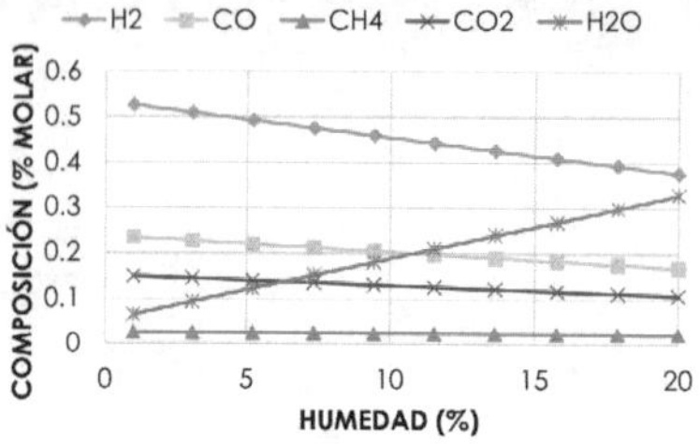

Coffee Cut-Stems. Effect of moisture in the Reduction Zone

Pinus Patula. Effect of moisture in the Reduction Zone

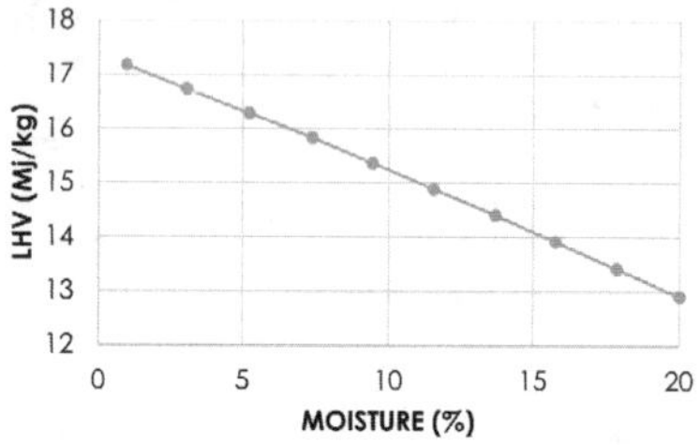

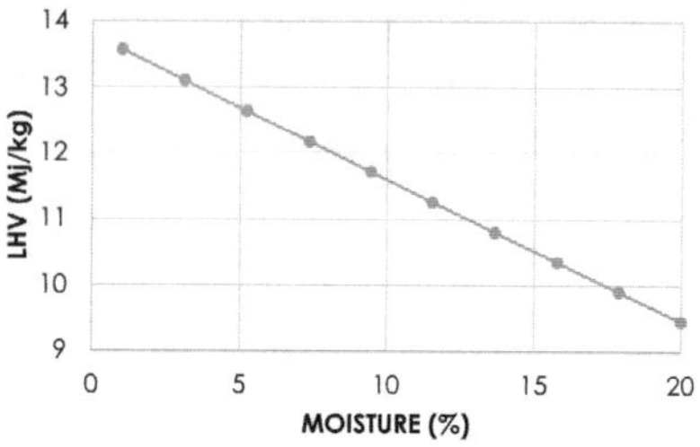

Chemical, Catalitic and Biotechnological Processes Research Group 15

Effect of Air/Biomass Ratio

3. Results

Pinus Patula

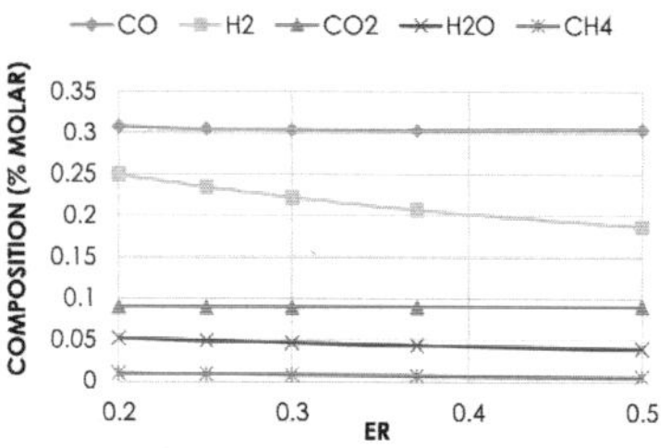

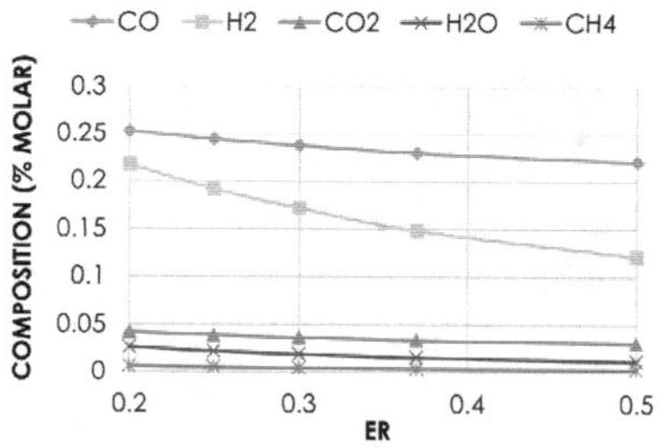

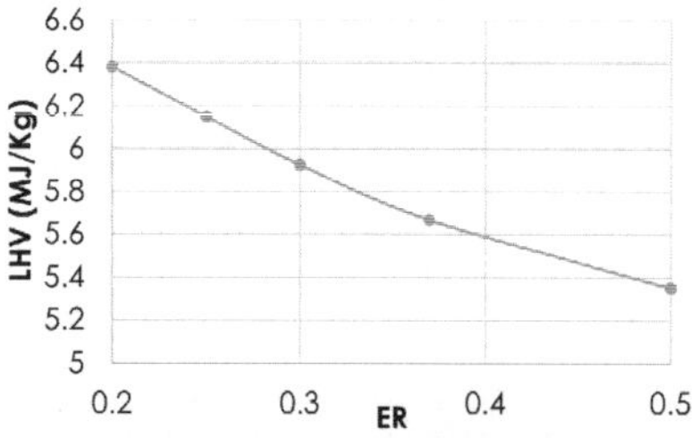

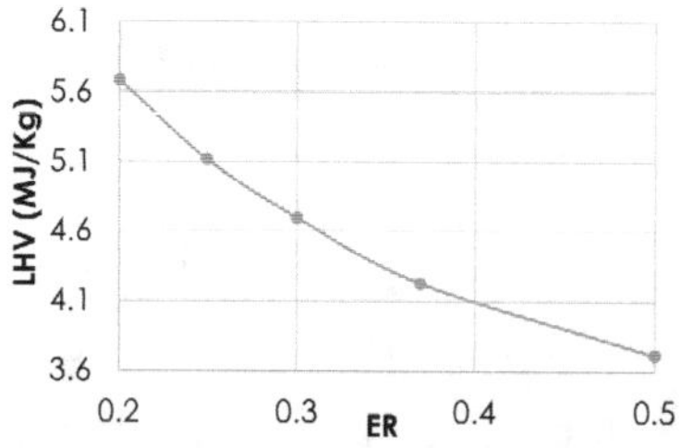

Coffee Cut-Stems

Chemical, Catalitic and Biotechnological Processes Research Group 16

Energy and Exergy

3. Results

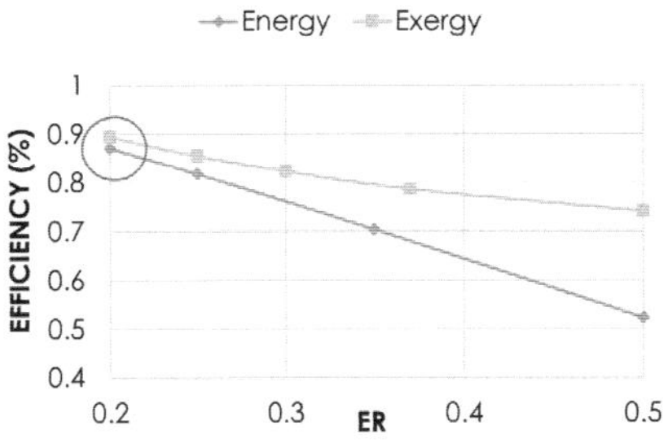

Pinus Patula

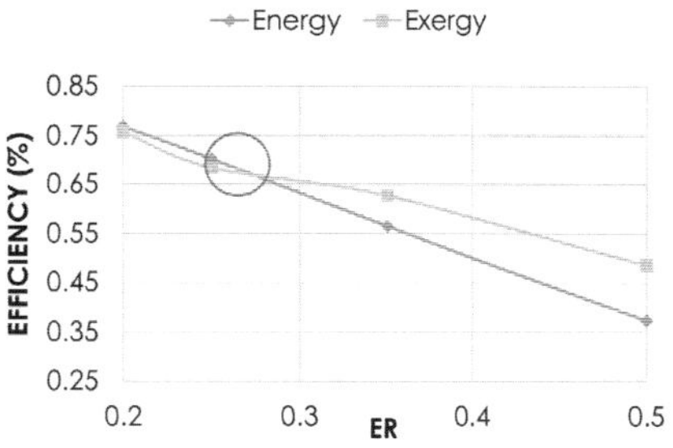

Coffee Cut-Stems

Simulation Results

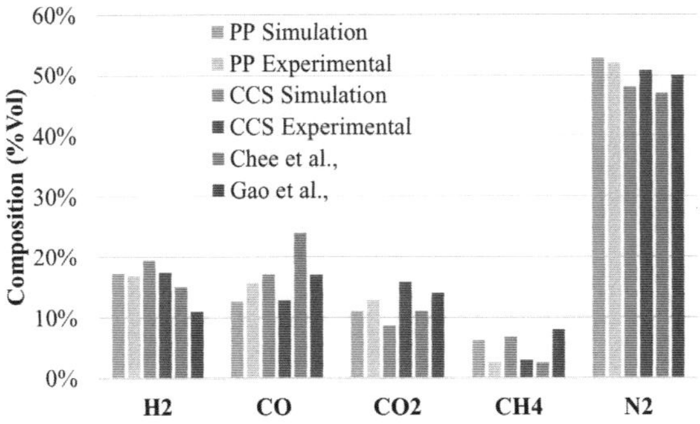

Gao. N, Li. A, Modeling and simulation of combied pyrolysis and reduction zone for a downdraft biomass gasifier, **Energy Conversion and Management**, *2008, pp. 3483 – 3490.*
Chee CS, The air gasification of Wood chips in a downdraft gasifier, **MSC Thesis, Kansas State University**, *1987.*

Raw Material	Process Yield	Units
Pinus Patula	6.26	Mj/kg
Coffee Cut-Stems	7.25	Mj/kg

Economic Assessment

3. Results

Raw Material Capacity vs. Production Cost

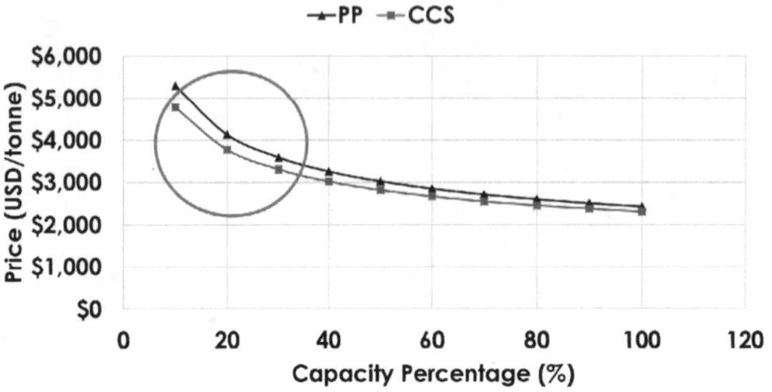

Hydrogen Market Price
4,740 USD/tonne

McKinsey. (2010). A portfolio of power-trains for Europe: a fact-based analysis - The role of Battery Electric Vehicles, Plug-in Hybrids and Fuel Cell Electric Vehicles. *Fuel Cell*, 68.

Environmental Assessment

3. Results

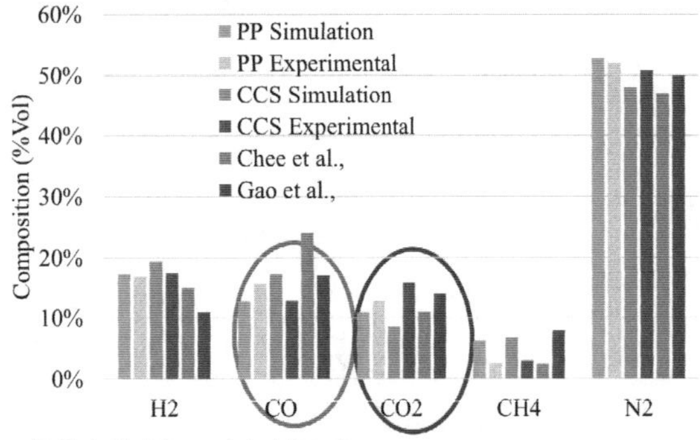

Gao. N, Li. A, *Modeling and simulation of combied pyrolysis and reduction zone for a downdraft biomass gasifier*, ***Energy Conversion and Management***, 2008, pp. 3483 – 3490.
Chee CS, *The air gasification of Wood chips in a downdraft gasifier*, **MSC Thesis, Kansas State University**, 1987.

Emission Factors
CO_2 = 1 kg/kg CO_2
CO = 1.57 kg/kg CO_2

GHG Balance

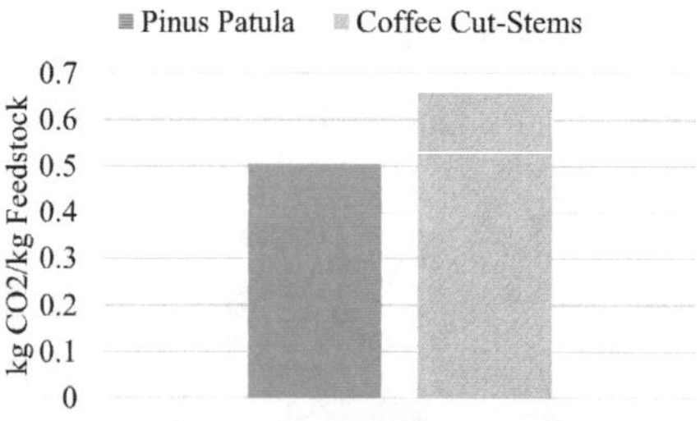

Conclusions

- Kinetic and equilibrium models properly describe the behavior of the biomass gasification using a simulation approach.
- The moisture content of the raw material affects the energy needs of the process, meanwhile the air/biomass ratio has direct influence in the synthesis gas composition.
- Process scale and the amount of available woody materials has direct effect on the hydrogen production cost, increasing the profitability of the process. However, logistic and transportation issues affect the performance of the process.

Chemical, Catalitic and Biotechnological Processes Research Group 21

Acknowledgements

The authors of this study acknowledge to the National University of Colombia at Manizales, the 703 announcement (Jóvenes Investigadores) of COLCIENCIAS and the Research Direction of Manizales (DIMA) for the financial support to develop this work.

22

References

Abuadala, a., Dincer, I., & Naterer, G. F. (2010). Exergy analysis of hydrogen production from biomass gasification. *International Journal of Hydrogen Energy, 35*(10), 4981–4990.

Dejtrakulwong, C., & Patumsawad, S. (2014). Four Zones Modeling of the Downdraft Biomass Gasification Process: Effects of Moisture Content and Air to Fuel Ratio. *Energy Procedia, 52*, 142–149.

Gao, N., & Li, A. (2008). Modeling and simulation of combined pyrolysis and reduction zone for a downdraft biomass gasifier. *Energy Conversion and Management, 49*(12), 3483–3490.

Jarungthammachote, S., & Dutta, a. (2007). Thermodynamic equilibrium model and second law analysis of a downdraft waste gasifier. *Energy, 32*, 1660–1669.

Future Work

Effect of the supply chain of the lignocellulosic residues (Pinus Patula and CCS) in the economic assessment of the hydrogen production.

Cradle to Gate Life Cycle Assessment of the Pinus Patula in Colombia for hydrogen production.

Chemical, Catalitic and Biotechnological Processes Research Group 23

Mathematical and Experimental Modelling of Biomass Gasification for Hydrogen Production

15AIChE
Annual Meeting, Salt Lake City, UT

Carlos A. García, Álvaro Gómez, Pisarenko Yu. A,
Carlos A. Cardona
e-mail: caagarciave@unal.edu.co
Universidad Nacional de Colombia. Sede Manizales

Chemical, Catalitic and Biotechnological Processes Research Group 24

423763 Multi-Fuel Scaled-Down Pure Hydrogen Generator: Design and Proof of Concept

Abstract:

Multi-fuel scaled-down autonomous pure H_2 generator:

Design and proof of concept

Michael Patrascu*, Moshe Sheintuch

Technion-Israel Institute of Technology, Department of Chemical Engineering, Haifa 32000, Israel

The world is showing an increasing interest in H_2 production for PEM fuel cell owing to the breakthroughs in fuel cell technology in the late 1990s. Conventionally H_2 is produced mainly through methane steam reforming (MSR), conducted in packed bed tubes filled with Ni based catalyst which are placed in a furnace supplying the heat of reaction through the burning of fuel. Full conversion of MSR to syngas (mixture of CO and H_2) is obtained at typical operation conditions of ~25 bar and temperatures as high as 1000°C. This unit is followed by high and low temperature water-gas-shift (WGS) to produce as much H_2 as possible. These steps are followed by pressure swing adsorption (PSA) for CO_2 removal and methanation (reverse steam reforming) for clearing any residual CO. Adopting this large scale production (~10^5 Nm^3/h) scheme as a H_2 source for electrical power production in PEMFC holds significant complications concerning compression, transportation and storage of H_2. Special infrastructure would be required to supply pure H_2 to a small-scale end user, either stationary or portable applications, with major safety and economic concern. Thus alternate process schemes for local scaled-down production of H_2 from conventional, environmentally friendly and safe feedstocks (e.g. natural gas, biofuels) are required. Alternative sources for H_2 production are desired, with lower (or zero) related carbon emissions, such as biofuels; methanol, ethanol or glycerol (which is a biodiesel production by-product) etc.

In this paper we successfully demonstrate, for the first time to our best knowledge, the feasibility of producing pure atmospheric pressure H_2, in a compact integrated membrane reactor, from various fuels. Demonstrating pure atmospheric H_2 production in one unit is challenging, and most previous studies used either vacuum or sweep gas to facilitate H_2 permeation in membrane reactors. Here this is achieved by steam reforming reaction using a commercial Ni catalyst and Pd-Ag membranes for *in situ* separation of the H_2 product. Feeds of methane, ethanol or glycerol and stoichiometric steam are considered, with the heat being supplied by recycling the SR effluents to an internal combustion reactor (Figure 1). The operating temperature in this auto-thermal mode of operation is not controllable independently, and is directly set by the feed flow rate. The production system's performance is characterized in terms of thermal efficiency of conversion (up to 25%) and power output (up to 0.15kW), Figure 2. Changing the fuel source leads to similar qualitative behavior, with small quantitative differences, mainly the operating temperature obtained. The feasibility of operating the same unit with various fuel types holds great potential and offers flexibility for technology developments based on H_2 as energy vector. Efficiencies obtained with MSR were better than those with ethanol SR or glycerol SR when compared at same maximal temperatures. Still these efficiencies are about 25% at best. Noting that around 50% of the energy is lost to the surroundings, we expect that increasing membrane area and applying better insulation will push the efficiencies to ~60%. A mathematical model predicts the results and operational trends well, while accounting on membrane permeance inhibition, which is significant, and affects the results and efficiencies.

Of the three feeds tested here methane is the best, having the largest H/C ratio, but ethanol, glycerol and other biofuels can be delivered in liquid phase, mixed with water. Other fuels like methanol can also be considered; methanol will allow to reduce the operating temperature to around 300°C. The SR and WGS catalytic reactions are fast and the power density is limited by the membrane permeance. With thinner membrane and better permeance it will be possible to improve the power density. Moreover, in many studies the permeance was found to be inhibited by co-adsorption of CO and methane on the Pd-Ag membrane. As we show here, accounting for such inhibition leads to better prediction of the experimental results.

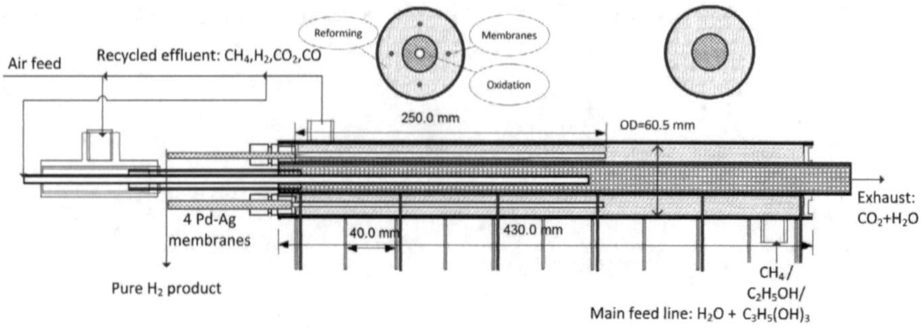

Figure 1: A detailed drawing of the reactor. The cross sections of the two reactor zones are depicted above. The 11 outer and 6 inner thermocouples (greed) and 2 of the 4 membranes (red in the colored version) are depicted in the axial view. The recycled SR effluents are fed to the internal combustion reactor through 2 axial locations using the branch tee (left) and an inner hasteloy tube.

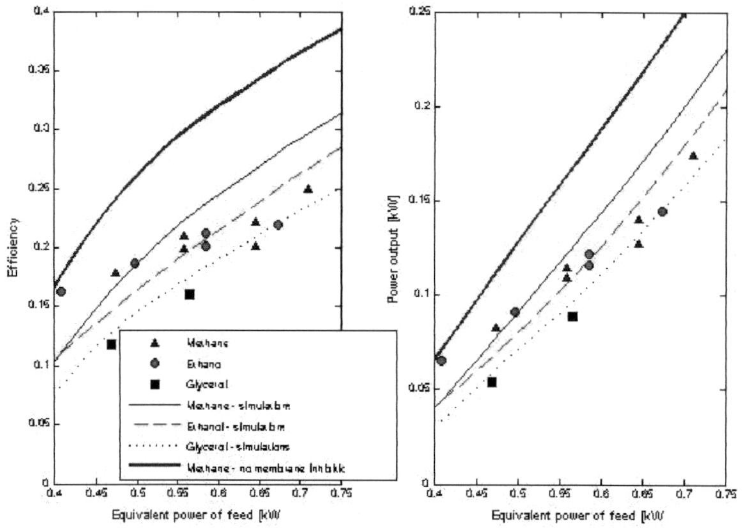

Figure 2: System's performance for different fuels fed vs the inlet fuel's equivalent power: (a) thermal efficiency and (b) equivalent power of H_2 product. Markers are experimental measurements and lines are model predictions. Stoichiometric feed (S/M=2 S/E=S/G=3), pressure of 11 bar.

15AIChE
Annual Meeting, Salt Lake City, UT

Solar Fuel Production via GeO_2/Geo Thermochemical H_2O and CO_2 Splitting Cycle

(Principal Author)
Rahul Bhosale
Assistant Professor
Department of Chemical Engineering
College of Engineering
Qatar University, Doha, Qatar
rahul.bhosale@qu.edu.qa

(Other Authors)
Anand Kumar
Assistant Professor
Department of Chemical Engineering
College of Engineering
Qatar University, Doha, Qatar
akumar@qu.edu.qa

Fares A. AlMomani
Associate Professor
Department of Chemical Engineering
College of Engineering
Qatar University, Doha, Qatar
falmomani@qu.edu.qa

Moustafa Hussein Ali
Undergraduate Student
Department of Chemical Engineering
College of Engineering
Qatar University, Doha, Qatar
ma1203561@student.qu.edu.qa

Mehak Jilani
Undergraduate Student
Department of Chemical Engineering
College of Engineering
Qatar University, Doha, Qatar
mj1104333@student.qu.edu.qa

Fatima Al-Massih
Undergraduate Student
Department of Chemical Engineering
College of Engineering
Qatar University, Doha, Qatar
fa1002721@student.qu.edu.qa

Dareen Dardor
Undergraduate Student
Department of Chemical Engineering
College of Engineering
Qatar University, Doha, Qatar
sg1000354@student.qu.edu.qa

Noor Alemadadhi
Undergraduate Student
Department of Chemical Engineering
College of Engineering
Qatar University, Doha, Qatar
na093583@student.qu.edu.qa

Roghayeh Dejan
Undergraduate Student
Department of Chemical Engineering
College of Engineering
Qatar University, Doha, Qatar
rd1105246@student.qu.edu.qa

Prepared for Presentation at
American Institute of Chemical Engineers
2015 Annual Meeting
Salt Lake City, Utah
November 8-13, 2015

Solar Fuel Production via GeO_2/Geo Thermochemical H_2O and CO_2 Splitting Cycle

Rahul Bhosale
Department of Chemical Engineering
College of Engineering
Qatar University, Doha, Qatar
rahul.bhosale@qu.edu.qa

Anand Kumar, Fares A. AlMomani, Moustafa Hussein Ali, Mehak Jilani, Fatima Al-Massih, Dareen Dardor, Noor Alemadadhi, Roghayeh Dejan

Department of Chemical Engineering
College of Engineering
Qatar University, Doha, Qatar

Keywords: Germanium Oxide, Water Splitting, CO_2 Splitting, Solar Fuel, Computational Thermodynamic Modeling

Abstract

This paper reports the thermodynamic feasibility of the production of solar fuel via GeO_2/GeO Thermochemical H_2O and CO_2 Splitting Cycle. This is a two-step process in which the first step is solar thermal dissociation of GeO_2 into GeO and the second step is non-solar splitting of H_2O and CO_2 via oxidation of GeO into GeO_2. Computational thermodynamic HSC Chemistry software and databases were used for the thermodynamic modeling of this process. This paper includes chemical thermodynamic equibrium analysis and second law thermodynamic modeling of the GeO_2/GeO based thermochemical CO_2 splitting process. Results obtained via the computational thermodynamic modeling are presented.

1. Introduction

As the world population continues to grow, the world has become obsessed with the shortage of fossil fuels and the main challenge for most countries is securing fuel supplies. Apart from the fuel demands, the problem of CO_2 emission is another global issue concerning all societies. CO_2 is the main reason causing the greenhouse effect, which leads to global warming and other consequences such as glaciers melting, sea level rise, and extreme climates, etc. A two-step solar thermochemical H_2O and/or CO_2 splitting process which utilizes metal oxide (MO) based redox reactions is one of the most promising ways for producing renewable fuels such as C, CO, H_2 or even a combination of CO and H_2 gas which is called syngas. In previous investigations, in addition to other H_2 production techniques, various MO based redox systems were investigated for the solar thermochemical H_2O and/or CO_2 splitting such as ZnO/Zn, SnO_2/SnO, Fe_3O_4/FeO, mixed ferrites, and ceria based oxides [1-62].

Note: Do not add page numbers. Do not refer to page numbers when referencing different portions of the paper

In this investigation, thermodynamic analysis of the GeO_2/GeO based solar thermochemical CO_2 splitting process was investigated in two sections with the help of commercial thermodynamic HSC Chemistry software and databases [37]. Firstly, chemical thermodynamic equilibrium of the GeO_2/GeO based solar thermochemical CO_2 splitting cycle was studied by examining the effect of various operating parameters on the thermal reduction and CO_2 splitting temperature. In second section, the second law thermodynamic analysis of this process was performed. The solar absorption efficiency of the solar reactor ($\eta_{absorption}$), solar energy input to the solar reactor (Q_{solar}), net energy absorbed in the solar reactor ($Q_{reactor-net}$), rediation heat losses from the solar reactor ($Q_{re-radiation}$), rate of heat rejected to the surrounding from the quench unit (Q_{quench}), and from the GeO oxidizer ($Q_{GeO-Oxd}$) , and solar to chemical energy conversion efficiency ($\eta_{solar-to-chemical}$) were determined by performing the thermodynamic modeling and are presented in this paper. A typical GeO_2/GeO based two-step solar thermochemical CO_2 splitting process is presented in Figure 1.

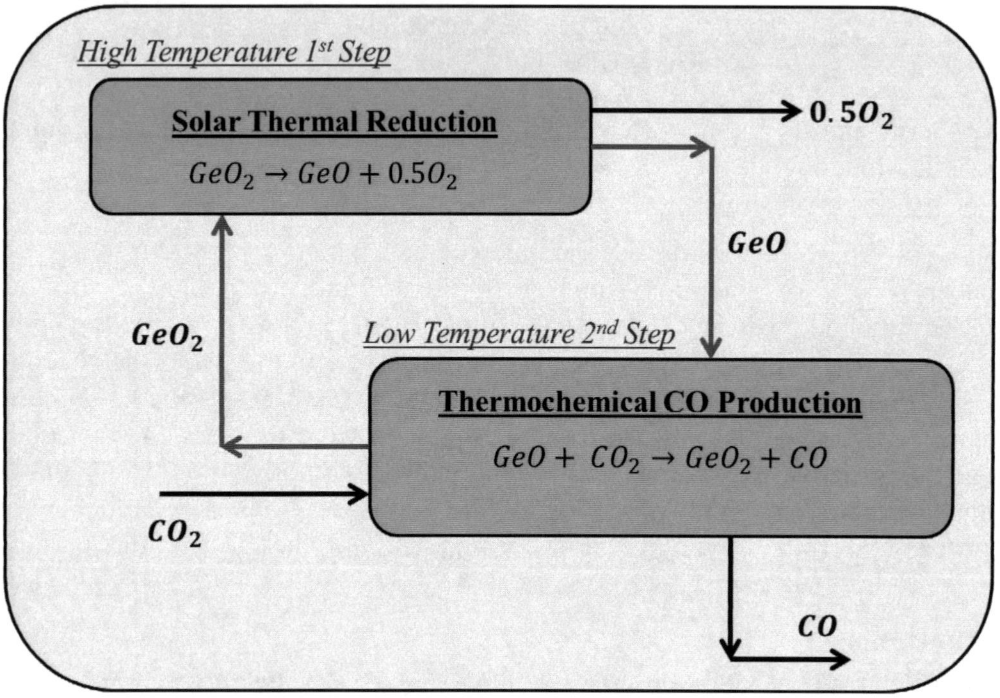

Figure 1. GeO_2/GeO based solar thermochemical CO_2 splitting process.

2. Chemical Thermodynamic Equilibrium

As per Figure 2, the solar thermal reduction of GeO_2 into GeO and $O_2(g)$ is possible at 2523 K (in absence on inert Ar). The solar thermal reduction reaction in case of GeO_2/GeO can be represented as follows:

$$GeO_2 \rightarrow GeO + {}^1/_2\, O_2(g) \tag{1}$$

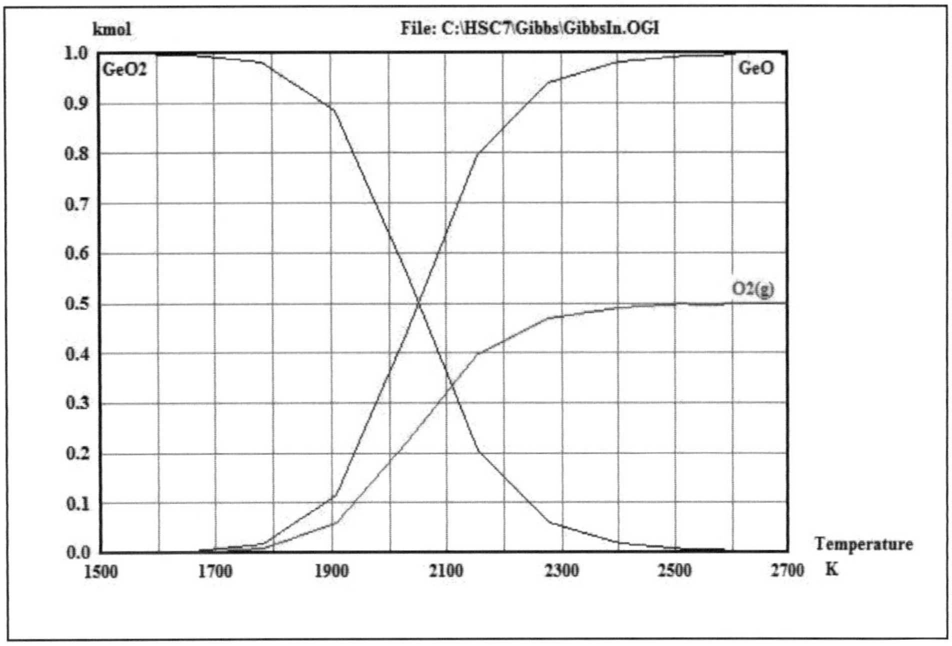

Figure 2. Solar thermal reduction of GeO_2 (Ar = 0 mol/sec).

Similarly, Figure 3 shows the progress of the oxidation of GeO into GeO_2 via water splitting reaction (in absence of Ar). According to the reported results the production of H_2 is feasible below 1340 K (according to reaction 2).

$$GeO + CO_2(g) \rightarrow GeO_2 + CO(g) \tag{2}$$

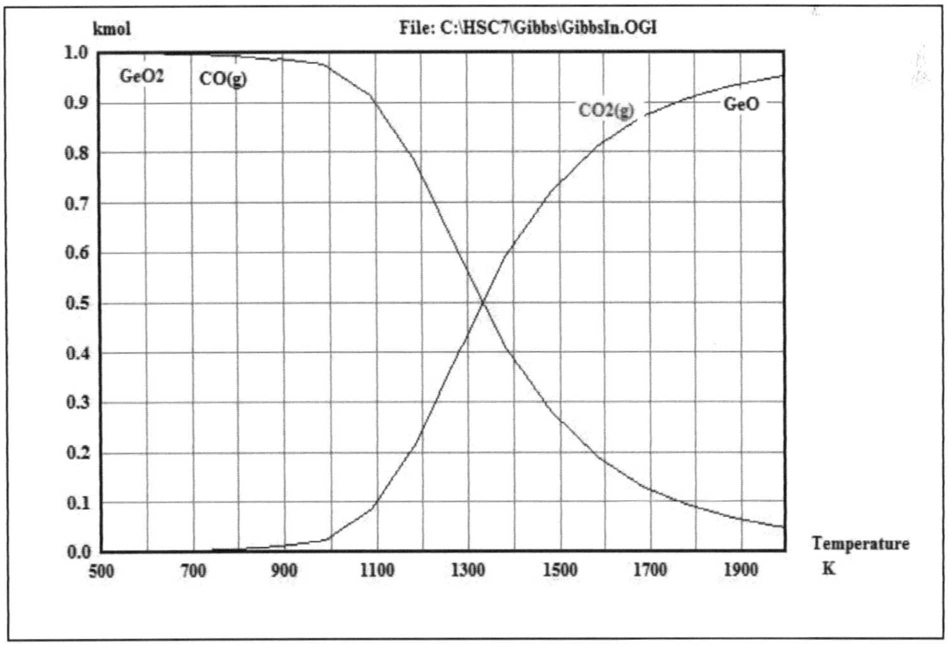

Figure 3. CO_2 splitting via GeO_2/GeO based solar thermochemical cycle.

3. Second Law Exergy Analysis

After determining the thermodynamic equilibrium compositions, the solar reactor efficiency of the GeO_2/GeO based solar thermochemical CO_2 splitting process is estimated by performing second law exergy analysis. The continuous GeO_2/GeO based CO production process consists of a solar reactor, a quench unit, a CO_2 splitting unit, a gas separator and a theoretical CO/O_2 ideal fuel cell. The process is assumed to operate under steady state conditions at a total pressure of 1 atm. Some aspects have been omitted from the calculations these include heat exchangers used for recovering sensible latent heat. Furthermore, assumptions include: 1) perfectly insulated solar reactor 2) no convective/conductive heat losses, 3) kinetic and potential energies are neglected, 3) all reactions proceed to completion. Thermodynamic properties are extracted from HSC software and databases and the analysis is normalized to the GeO_2 molar flow rate (1 mol/sec) entering the solar reactor.

Solar energy absorption efficiency ($\eta_{absorption}$)

Solar energy absorption efficiency is defined as the ratio of the net rate at which energy is being absorbed by the solar reactor divided by the solar power input through the aperture of the solar reactor and can be calculated using the following formula:

$$\eta_{absorption} = \frac{Q_{reactor-net}}{Q_{solar}} = 1 - \left(\frac{\sigma T_H^4}{IC}\right) \quad (3)$$

Where, σ = Stefan-Boltzmann constant, I = normal beam insolation $1000(W/m^2)$, C= solar flux concentration ratio of the solar concentrating system (5000 suns), $Q_{reactor-net}$= net energy absorbed in the solar reactor, Q_{solar} = solar energy input. For this process at thermal reduction temperature equal to 2523 the $\eta_{absorption}$ = 54.05%.

Net solar energy absorbed by the solar reactor ($Q_{reactor-net}$)

As per the process flow diagram for GeO_2/GeO based CO production process the continuous GeO_2 stream is heated up to the thermal reduction temperature. The amount of energy required for the complete thermal reduction of GeO_2 into GeO can be calculated as follows:

$$Q_{reactor-net} = Q_{GeO_2-heating} + Q_{GeO_2-reduction} \quad (4)$$

As per the performed computational thermodynamic modeling, the total amount of energy required for the solar thermal reduction of GeO_2 is equal to 487 kW.

Solar energy input to the solar reactor (Q_{solar})

For the obtained $\eta_{absorption}$ and $Q_{reactor-net}$, the total solar energy input needed to operate the GeO_2/GeO based CO production process can be determined by equation (5) and observed to be equal to 900.22 kW.

$$Q_{solar} = \frac{Q_{reactor-net}}{\eta_{absorption}}$$

Radiation heat losses from the solar reactor ($Q_{re-radiation}$)

Radiation heat losses from the solar reactor conducting the thermal reduction of GeO_2 can be calculated by using equation (6):

$$Q_{re-radiation} = Q_{Solar} - Q_{reactor-net} \tag{6}$$

Calculated results indicate radiation heat losses equal to 413.68 kW.

Rate of heat rejected to the surrounding from the coolers

In GeO_2/GeO based CO production process, three coolers are included in the process flow diagram. Cooler – 1 is used to cool the reduction products from reduction temperature to CO_2 splitting temperature (500 K). The amount of heat rejected by cooler – 1 equal to 114.15 kW. Cooler – 2 is used to cool the O_2 from 500K to room temperature and the amount of heat rejected equal to 3.14 kW. Also cooler – 3 is installed to cool the CO produced from 500K to room temperature and the heat rejected to surrounding is equal to 6.02 kW.

Rate of heat rejected to the surrounding by the CO_2 splitting reactor

Produced GeO from the solar reactor is first cooled down to 500K (CO2 splitting temperature) and then transferred to the CO_2 splitting reactor. It is noteworthy that the heat released from the exothermic CO_2 splitting reaction is sufficient enough to overcome the sensible and latent heats and raise the temperature to the required splitting temperature. For this step, complete conversion of GeO to GeO_2 is assumed and the rate of heat rejected to the surrounding from the CO2 splitting is estimated to be 85.54 kW.

$$Q_{Zn-oxd} = -\dot{n}\Delta H|_{\mathrm{GeO}(s)+CO_2(g)\rightarrow \mathrm{GeO}_2(s)+CO(g)@500\mathrm{K}} \tag{7}$$

Rates of theoretical work performed ($W_{FC-Ideal}$) ***and heat rejected*** ($Q_{FC-Ideal}$) ***by the fuel cell***

An ideal fuel cell is added in order to complete the GeO_2/GeO based CO production process and to calculate the maximum possible theoretical work that can be extracted from this cycle. The ideal fuel cell operates in the presence of CO/O_2, and the rates of theoretical work performed ($W_{FC-Ideal}$) and the heat rejected by the fuel cell ($Q_{FC-Ideal}$) can be calculated by using equations (8) and (9) where the efficiency of the fuel cell (η_{FC}) is assumed to be 100%.

$$W_{FC-Ideal} = -(\eta_{FC}) \times \dot{n}\,\Delta G|_{CO(g)+0.5O_2(g)\rightarrow CO_2(g)@298K} \tag{8}$$

$$Q_{FC-Ideal} = -(T_R) \times \dot{n}\,\Delta S|_{CO(g)+0.5O_2(g)\rightarrow CO_2(g)@298K} \tag{9}$$

For the GeO_2/GeO based CO production process, the $W_{FC-Ideal}$ and $Q_{FC-Ideal}$ are observed to be equal to 256.95 kW and 25.77 kW, respectively.

Cycle and solar-to-fuel conversion efficiency

Cycle efficiency of the solar thermochemical GeO_2/GeO based CO_2 splitting cycle is defined as the ratio of theoretical work performed by the ideal fuel cell to the solar energy input:

$$\eta_{cycle} = \frac{W_{FC-Ideal}}{Q_{solar}} \tag{10}$$

Furthermore, the solar-to-fuel energy conversion efficiency of the solar thermochemical GeO_2/GeO based CO_2 splitting cycle is defined as the ratio of heating value (HV) of the CO produced to the solar energy input:

$$\eta_{solar-to-fuel} = \frac{HV_{CO}}{Q_{solar}} \tag{11}$$

As per the HSC computation thermodynamic modeling, η_{cycle} and $\eta_{solar-to-fuel}$ equal to 28.54% and 31.43%, respectively.

4. Conclusion

This study reports the thermodynamic investigation of two-step solar thermochemical cycles for CO production via GeO_2/GeO based CO_2 splitting reactions. According to the performed calculations, it is observed that the solar thermal reduction of GeO_2 (100% conversion) is possible at 2523 K (in absence of Ar) whereas the CO2 splitting is feasible below 1340 K. The second law thermodynamic analysis of the solar reactor conducting the thermochemical GeO_2/GeO based CO_2 splitting process at 2523 K and at a solar concentration of 5000 suns indicate $\eta_{absorption}$ = 54.05%, Q_{solar}= 900.22 kW, $Q_{reactor-net}$ = 486.54 kW, $Q_{re-radiation}$ = 413.68 kW, $Q_{CO2-splitting}$ =85.54 kW, and $\eta_{solar-to-fuel}$ = 31.43%.

5. Acknowledgments

The authors gratefully acknowledge the financial support provided by the Qatar University Internal Grants QUUG-CENG-CHE-13/14-4 and QUUG-CENG-CHE-14\15-10.

References

[1] Olivier JGJ, Janssens-Maenhout G, A.H.W. Peters J. Trends in global CO_2 emissions 2012; 2012.

[2] Fakhrai, R., Saadatfar, B., & Wei, B., The thermodynamic analysis of two-step conversions of CO_2/H_2O for syngas production by ceria. International Journal of Hydrogen Energy, 39 (2014) 12353.

[3] Steinfeld A, Weimer AW. Thermochemical production of fuels with concentrated solar energy. Optics Express, 18 (2010) A100.

[4] Steinfeld A. Solar thermochemical production of hydrogen: a review. Solar Energy, 78 (2005) 603.

[5] Nikulshina V, Galvez ME, Steinfeld A. Kinetic analysis of the carbonation reactions for the capture of CO_2 from air via the $Ca(OH)_2$–$CaCO_3$–CaO solar thermochemical cycle. Chemical Engineering Journal, 129 (2007) 75.

[6] Nikulshina V, Gebald C, Steinfeld A. CO_2 capture from atmospheric air via consecutive CaO-carbonation and $CaCO_3$-calcination cycles in a fluidized-bed solar reactor. Chemical Engineering Journal, 146 (2009) 244.

[7] Nikulshina V, Hirsch D, Mazzotti M, Steinfeld A. CO_2 capture from air and co-production of H_2 via the $Ca(OH)_2$–$CaCO_3$ cycle using concentrated solar power-thermodynamic analysis. Energy, 31 (2006) 1715.

[9] Nikulshina V, Steinfeld A. CO_2 capture from air via CaO carbonation using a solar-driven fluidized bed reactor effect of temperature and water vapor concentration. Chemical Engineering Journal, 155 (2009) 867.

[10] Davidson, J., & Krenzke, P. Thermodynamic Analysis of Syngas Production via the Solar Thermochemical Cerium Oxide Redox Cycle with Methane-Driven Reduction. Energy & Fuels, 28 (2014) 4088.

[11] Loutzenhiser, P., Steinfeld, A. Solar syngas production from CO_2 and H_2O in a two-Step Thermochemical Cycle via Zn/ZnO Redox Reactions: Thermodynamic cycle Analysis. International Journal of Hydrogen Energy, 36 (2011) 12141.

[12] Stamatiou, A., Loutzenhiser, P., Steinfeld, A. Solar Syngas Production from H_2O and CO_2 via two-Step Thermochemical Cycles based on Zn/ZnO and FeO/Fe_3O_4 Redox Reactions: Kinetic Analysis. Energy & Fuels, 24 (2010) 2716.

[13] Charvin, P., Abanades, S., Lemont, F., Flamant, G. Experimental study of SnO_2/SnO/Sn thermochemical systems for solar production of hydrogen. AIChE Journal, 54 (2008) 2759.

[14] Abanades, S. CO_2 and H_2O reduction by solar thermochemical looping using SnO_2/SnO redox reactions: Thermogravimetric analysis. International Journal of Hydrogen Energy, 37 (2012) 8223.

[15] Chambon, M., Abanades, S., Flamant, G. Kinetic investigation of hydrogen generation from hydrolysis of SnO and Zn solar nanopowders. International Journal of Hydrogen Energy, 34 (2009) 5326.

[16] Vishnevetsky, I., Epstein, M., J. Tin as possible candidate for solar thermochemical redox process for hydrogen production. Journal of Solar Energy Engineering, 131 (2009) 021007-1.

[18] Abanades, S., Villafan-Vidales, H. CO_2 and H_2O conversion to solar fuels via two-step solar thermochemical looping using iron oxide redox pair. Chemical Engineering Journal 175 (2011) 368.

[19] Gokon, N., Murayama, H., Nagasaki, A., Kodama, T. Thermochemical two-step water splitting cycles by monoclinic ZrO_2 supported $NiFe_2O_4$ and Fe_3O_4 powders and ceramic foam. Solar Energy, 83 (2009) 527.

[20] Scheffe, J., Li, J., Weimer, A. A spinel ferrite/hercynite water-splitting redox cycle International Journal of Hydrogen Energy, 35 (2010) 3333.

[21] Roeb, M., Gathmann, N., Neises, M., Sattler, C., Pitz-Paal, R. Thermodynamic analysis of two-step solar water splitting with mixed iron oxides. International Journal of Energy Research, 33 (2009) 893.

[22] Lorentzou, S., Agrafiotis, A., Konstandopoulos, A., Aerosol spray pyrolysis synthesis of water-splitting ferrites for solar hydrogen production. Granular Matter, 10 (2008) 113.
[23] Abanades, S., Flamant, G. Thermochemical hydrogen production from a two-step solar-driven water-splitting cycle based on cerium oxides. Solar Energy, 80 (2006) 1611.
[24] Chueh, W. C., Falter, C., Abbott, M., Scipio, D., Furler, P., Haile, S., Steinfeld, A. High-Flux Solar-Driven Thermochemical Dissociation of CO_2 and H_2O using Nonstoichiometric Ceria. Science, 330 (2010) 1797.
[25] Kaneko, H., Miura, T., Ishihara, T., Taku, S., Yokoyama, T., Nakajima, H., Tamaura, Y. Cerium ion redox system in CeO2-xFe2O3 solid solution at high temperatures (1273-1673K) in the two-step water splitting reaction for solar H_2 generation. Energy, 43 (2008) 3153.
[26] Meng, Q., Lee, C., Ishihara, T., Kaneko, H., Tamaura, Y. Reactivity of CeO2-based ceramics for solar hydrogen production via a two-step water-splitting cycle with concentrated solar energy. International Journal of Hydrogen Energy, 36 (2011) 13435.
[27] Le Gal, A., Abanades, S. Catalytic investigation of ceria-zirconia solid solutions for solar hydrogen production. International Journal of Hydrogen Energy, 36 (2011) 4739.
[28] Abanades, S., Gal, A. L. CO_2 Splitting by Thermo-Chemical Looping Based on $Zr_xCe_{1-x}O_2$ Oxygen Carriers for Synthetic Fuel Generation. Fuel, 102 (2012) 180.
[29] Bhosale RR, Shende RV, Puszynski JA, H2 generation from thermochemical water-splitting using sol-gel derived Ni-ferrite. Journal of Energy and Power Engineering 2010;4:27–38.
[30] Bhosale RR, Dardor D, Gharbia S, Folady J, Jilani M, Kumar A, van den Broke LJP, Lin F, Alexneit I, Thermochemical Conversion of CO_2 into Solar Fuels Using Ferrite Nanomaterials. In: Proceedings of 4th International Gas Processing Symposium 2015;141-185.
[31] Bhosale RR, Yousefi S, Folady J, Jilani M, Dardor D, Gharbia S, Kumar A, van den Broke LJP, Alexneit I, Shende R, Thermodynamic Analysis of Solar Fuel Production via Thermochemical H_2O and/or CO_2 Splitting Using Tin Oxide Based Redox Reactions. In: Proceedings of 4th International Gas Processing Symposium 2015;39 – 48.
[32] Bhosale RR, Jilani M, Dardor D, Gharbia S, Folady J, Kumar A, van den Broke LJP, Lin F, Alexneit I, Solar Fuel Production via Non-Stoichiometric $Ce_xZr_yHf_zO_{2-\delta}$ Based Two-Step Thermochemical Redox Cycle. In: Proceedings of 4th International Gas Processing Symposium 2015;117 – 124.
[33] Kumar A, Ashok A, Kuti YO, Bhosale RR, van den Broke LJP, Single Step Synthesis of Transition Metal Nanoparticles in Aqueous Phase for Catalytic Applications. In: Proceedings of 4th International Gas Processing Symposium 2015;69 – 80.
[35] Dowass NA, Alazewi NS, Shurair MS, Bhosale RR, van den Broke LJP, Kumar A, Modeling of reaction front movement in combustion synthesis for catalyst preparation. In: Proceedings of 4th International Gas Processing Symposium 2015;391 – 399.
[37] Dardor D, Bhosale RR, Gharbia S, Kumar A, AlMomani F, Solar Carbon Production via Thermochemical ZnO/Zn Carbon Dioxide Splitting Cycle. Journal of Emerging Trends in Engineering & Applied Sciences 2015;6:129 – 135.
[38] Bhosale RR, Kumar A, AlMomani F, Alexneit I, Propylene Oxide Assisted Sol-Gel Synthesis of Zinc Ferrite Nanoparticles for Solar Fuel Production. Ceramics International 2015 (in press).

[39] Dardor D, Bhosale RR, Gharbia S, Kumar A, AlMomani F, Solar Thermochemical Conversion of CO_2 into C via SnO_2/SnO Redox Cycle: Thermodynamic Study. International Journal of Engineering Research & Applications 2015;5:134 – 140.

[40] Bhosale RR, Kumar A, AlMomani F, Gharbia S, Dardor D, Ali M, Folady J, Yousefi S, Jilani M, Solar Thermochemical Reforming of Natural Gas via Metal Oxide Based Redox Reactions. In: Proceedings of 2015 AIChE Spring Meeting (ISBN: 978-0-8169-1089-2) 2015.

[41] Bhosale RR, Kumar A, AlMomani F, Gharbia S, Dardor D, Ali M, Folady J, Yousefi S, Jilani M, Al-Fakih M, A Novel Two-Step Metal Oxide – Metal Sulfate Water Splitting Cycle for the Production of Solar Hydrogen. In: Proceedings of 2015 AIChE Spring Meeting (ISBN: 978-0-8169-1089-2) 2015.

[42] Bhosale RR, Kumar A, AlMomani F, Dardor D, Gharbia S, Ali M, Folady J, Yousefi S, Jilani M, Abuarja D, Angre P, Solar Hydrogen Production via Thermochemical Metal Oxide – Metal Sulfate Water Splitting Cycle. In: Proceedings of 2015 AIChE Spring Meeting (ISBN: 978-0-8169-1089-2) 2015.

[43] Dardor D, Bhosale RR, Kumar A, AlMomani F, Gharbia S, Ali M, Folady J, Yousefi S, Jilani M, Computational Thermodynamic Analysis of Solar Fuel Production via Metal Oxide Based H_2O and CO_2 Splitting Thermochemical Cycles. In: Proceedings of 2015 AIChE Spring Meeting (ISBN: 978-0-8169-1089-2) 2015.

[44] Gharbia S, Bhosale RR, Kumar A, AlMomani F, Dardor D, Ali M, Folady J, Yousefi S, Jilani M, Al-Fakih M, Angre P, Thermochemical Conversion of H_2O and CO_2 into Solar Fuels via Metal Oxide Based Redox Reaction. In: Proceedings of 2015 AIChE Spring Meeting (ISBN: 978-0-8169-1089-2) 2015.

[45] Ashok A, Kumar A, Bhosale RR, Saleh MAH, van den Broeke LJP, Cellulose Assisted Combustion Synthesis of Porous Cu-Ni Nanopowders. RSC Advances 2015;5:28703 – 28712.

[46] Kumar A, Cross A, Manukyan K, Bhosale RR, van den Broeke LJP, Miller JT, Mukasyan AS, Wolf EE, Combustion Synthesis of Copper-Nickel Catalysts for Hydrogen Production from Ethanol. Chemical Engineering Journal 2015;278:46 – 54.

[47] Bhosale RR, Kumar A, van den Broeke LJP, Dardor D, Gharbia S, Jilani M, Folady J, Al-Fakih M, Tarsad M, Solar Hydrogen Production via Thermochemical Iron Oxide – Iron Sulfate Water Splitting Cycle. International Journal of Hydrogen Energy 2015;40:1639 – 1650.

[48] Bhosale RR, Alxneit I, Kumar A, van den Broeke LJP, Folady J, Dardor D, Gharbia S, Jilani M, Al-Fakih M, Elshrbajy NN, Tarsad M, Almkhyari FA, Solar Thermochemical Conversion of CO_2 into Synthetic Fuels via Ferrite Based Redox Reactions. In: Proceedings of 2014 AIChE Annual Meeting (ISBN: 978-0-8169-1086-1) 2014.

[49] Bhosale RR, Alxneit I, Kumar A, van den Broeke LJP, Jilani M, Folady J, Dardor D, Gharbia S, Dejan R, Al-Emadi N, Al-Massih F, Al-Yazedi S, Thermochemical Dissociation of CO_2 into Renewable Fuels via $Ce_xZr_yHf_zO_2$ Based Redox Reactions Using Concentrated Solar Energy. In: Proceedings of 2014 AIChE Annual Meeting (ISBN: 978-0-8169-1086-1) 2014.

[50] Kumar A, Ashok A, Kuty YO, Bhosale van den Broeke LJP, Combustion Synthesis of Copper-Nickel Catalysts for Hydrogen Production from Ethanol. In: Proceedings of 2014 AIChE Annual Meeting (ISBN: 978-0-8169-1086-1) 2014

[51] Bhosale RR, Alxneit I, van den Broeke LJP, Kumar A, Jilani M, Gharbia S, Folady J, Dardor D, Sol-Gel Synthesis of Nanocrystalline Ni-Ferrite and Co-Ferrite Redox Materials for Thermochemical Production of Solar Fuels. Materials Research Society Symposium Proceedings 2014;1675:mrss14-1675-rr06-10.

[52] Bhosale RR, Shende RV, Puszynski JA, Sol-Gel Derived $NiFe_2O_4$ Modified With ZrO_2 for Hydrogen Generation from Solar Thermochemical Water-Splitting Reaction. Materials Research Society Symposium Proceedings 2012;1387:mrsf11-1387-e09-07.

[53] Pasala X, Bhosale RR, Amar VS, Puszynski JA, Shende RV, Thermally Stabilized Ferrite Nanoparticles for Hydrogen Generation from Thermochemical Water-Splitting Reaction. In: Proceedings of 2012 AIChE Annual Meeting (ISBN 978-0-8169-1073-1) 2012.

[54] Bhosale RR, Pasala X, Yelakanti S, Puszynski JA, Shende RV, Solar Thermochemical Water-Splitting for H_2 Generation Using Sol-Gel Derived Ferrite Nanomaterials. In: Proceedings of NSTI – Nanotech Conference & Expo., (ISBN: 978-1-4665-6276-9) 2012;3:555.

[55] Bhosale RR, Shende RV, Puszynski JA, Synthesis of Nano-Sized Ferrite Materials for H2 Generation from High Temperature Water-Splitting Reaction. In: Proceedings of NSTI – Nanotech Conference & Expo., (ISBN: 978-1-4398-7142-3) 2011;1:237.

[56] Bhosale RR, Shende RV, Puszynski JA, H_2 Generation from Thermochemical Water-Splitting Using Sol-Gel Synthesized Ferrites ($M_xFe_yO_z$, M = Ni, Zn, Sn, Co, Mn, Ce). In: Proceedings of 2010 AIChE Annual Meeting (ISBN: 978-0-8169-1065-6) 2010.

[57] Bhosale RR, Shende RV, Puszynski JA, Sol-Gel Synthesis of Ferrite Foam Materials for H_2 Generation from Water-Splitting Reaction. In: Proceedings of NSTI – Nanotech Conference & Expo., (ISBN: 978-1-4398-3401-5) 2010;1:368.

[58] Bhosale RR, Shende RV, Opoku MK, Puszynski JA, H_2 Generation from Water-Splitting Using Novel Ferrite Foam Materials. In: Proceedings of 2009 AIChE Annual Meeting (ISBN: 978-0-8169-1058-6) 2009.

[59] Shende RV, Puszynski JA, Opoku MK, Bhosale RR, Synthesis of Novel Ferrite Foam Materials for Water-Splitting Application. In: Proceedings of NSTI – Nanotech Conference & Expo., (ISBN: 978-1-4398-1782-7) 2009;1:201.

[60] Bhosale RR, Shende RV, Puszynski JA, H2 generation from thermochemical water-splitting using sol-gel synthesized Zn/Sn/Mn-doped Ni-ferrite. International Review of Chemical Engineering 2010;2:852–62

[61] Bhosale RR, Khadka RP, Shende RV, Puszynski JA. H2 generation from two-step thermochemical water-splitting reaction using sol-gel derived SnxFeyOz. Journal of Renewable and Sustainable Energy 2011;3:063104-1–063104-16.

[62] Bhosale RR, Shende RV, Puszynski JA. Thermochemical water-splitting for H2 generation using sol-gel derived Mn-ferite in a packed bed reactor. International Journal of Hydrogen Energy 2012;37:2924–34.

417414 Extended Surface Electrocatalysts in the Oxygen Evolution Reaction

Abstract:

Although electrochemical water splitting currently represents a small amount of the US hydrogen supply, its role is expected to grow with the emergence of a hydrogen-based infrastructure.[1] Electrolysis is reasonably cost-competitive with a variety of technologies, including batteries (solid state and flow), pumped hydro, and compressed air.[2] When coupled with renewable energy sources, such as wind and solar, electrochemical water splitting can produce hydrogen using a nearly carbon free pathway.[3] The commercial progress of electrolyzers is limited in part by the cost of the catalyst layer. The oxygen evolution reaction is the focus of electrocatalyst development due to slow kinetics relative to hydrogen evolution; iridium nanoparticles are commonly used to catalyse oxygen evolution, with platinum added to ensure conductivity throughout the catalyst layer.

Extended surface nanostructures have previously been developed as electrocatalysts in hydrogen fuel cells, where they have produced an order of magnitude higher specific activity than nanoparticles in oxygen reduction and demonstrated a potential benefit in accelerated stress tests.[4] Extended nanostructures have recently been developed for the oxygen evolution reaction, where a similar specific activity improvement was observed in rotating disk electrode (RDE) half-cells. Catalysts formed by spontaneous galvanic displacement are ideally situated, being able to take advantage of the specific activities generally associated with the catalyst type, while significantly improving upon the traditionally low surface areas of extended surfaces. This approach has been used in forming iridium-nickel and iridium-cobalt nanowires.[5]

While nanoparticles have modest site-specific activity in oxygen reduction and evolution, nanoparticle evolution catalysts are at a particular disadvantage with regards to surface area, since carbon supports cannot be used at high potentials. Iridium-cobalt and iridium-nickel nanowires have been synthesized with surface areas of 70 m^2 g_{Ir} ^{1}, more than double the nanoparticles. In RDE half-cells, the extended nanostructures produce an oxygen evolution mass activity 7 to 8 times greater throughout the kinetic region. Accelerated stress tests were completed by potential holds and cycling over a variety of conditions to study catalyst degradation and establish standardized testing protocols. Acid treatment has been used to remove excess nickel and cobalt template, improving catalyst durability. Following accelerated stress tests in RDE half-cells, iridium-cobalt and iridium-nickel nanowires exceed the mass activity of nanoparticles by more than an order of magnitude.

Membrane electrode assembly (MEA) fabrication of these materials has produced electrolyzers with higher performance than the benchmark device. Recent improvements in MEA performance suggest that the promise of extended surface catalysts observed in half-cells can be realized in the application.

[1] R. Forgie, G. Bugosh, K.C. Neyerlin, Z. Liu, P. Strasser, Electrochemical and Solid-State Letters, 13 (2010) B36-B39.

[2] K. Harrison, M. Peters, in: U.S. Department of Energy (Ed.), http://www.hydrogen.energy.gov/pdfs/review13/pd031_harrison_2013_o.pdf, 2013.

[3] K. Harrison, M. Peters, C. Ainscough, in: U.S. Department of Energy (Ed.), http://www.hydrogen.energy.gov/pdfs/progress13/ii_a_2_harrison_2013.pdf, 2013.

[4] S.M. Alia, Y.S. Yan, B.S. Pivovar, Catalysis Science & Technology, 4 (2014) 3589-3600.

[5] H. Xu, High-Performance, Long-Lifetime Catalysts for Proton Exchange Membrane Electrolysis, in: U.S. Department of Energy (Ed.), http://www.hydrogen.energy.gov/pdfs/review14/pd103_xu_2014_o.pdf, 2014.

(720f) Kinetic Mechanisms of Hydrogen Sorption in Nanocrystalline Magnesium (hydride)

Shriniwasan, S., Indian Institute of Technology Bombay
Tien, H. Y., University of Florida, FL, USA
Tanniru, M., University of Florida
Ebrahimi, F., University of Florida
Gor, N., Indian Institute of Technology Bombay
Tatiparti, S. S., Indian Institute of Technology Bombay
Originally presented on: 11/12/2015 16:40:00 - 16:57:00

Abstract

Magnesium (hydride) is a very promising material for hydrogen storage with very high gravimetric capacity (7.6 wt.%). However, its slow kinetics of hydrogen absorption/desorption (sorption) hinders its application. Strategies such as particle size reduction, catalyst addition are employed to improve the kinetics of sorption. To device the strategies for improving the kinetics, it is essential to unravel the mechanisms of hydrogen sorption. The kinetics of hydrogen sorption in ball milled nanocrystalline magnesium (hydride) powders are investigated using the popular Johnson-Mehl-Avrami-Kolmogorov (JMAK) equation (α = 1?exp(?kt^n)) at different temperatures (180 °C-400 °C). The exponent n indicates the dimensionality of the growth of Mg (during desorption) or MgH_2 (during absorption). n decreases with time from a high value (>1) to a negligible value (<0.5) suggesting that the growth of the phase (Mg or MgH_2) decreases with time. The cross-sectional observation of these powders using SEM and TEM revealed that the depth of the growing phase from the particle surface increases with time rendering a core-shell structure during sorption. The estimated hydride/metal interface velocity (U) for growth of these phases shows two clear regimes of high and low velocities in the n>0.5 and n<0.5 regimes respectively. The former velocity is at least 2 orders of magnitude higher than the latter. The diffusion coefficients (D) are also estimated using Fick's 2nd law (transient) for hydrogen diffusion through the magnesium hydride phase. The estimated diffusion coefficients are at least 4 orders of magnitude higher in the n>0.5 regime than in those in the n<0.5 regime. Moreover, the diffusion coefficients in the n<0.5 regime are closer to the expected values from literature. The estimated activation energy for hydrogen diffusion through hydride is 91 kJ/mol H and is close to the literature reported value of 100±10 kJ/mol H. The combined n-U-D analysis supported by the microstructural examinations suggests that the growth of the magnesium or hydride is governed by hydride/metal interfacial movement in the n>0.5 regime and by hydrogen diffusion through hydride in the n<0.5 regime. The present analysis distinguishes growth by the interfacial velocity from that by diffusion during hydrogen sorption in magnesium (hydride). The present analysis contributes to the advancement of the understanding on the kinetic mechanisms of hydrogen sorption of MgH_2.

415004 Thermodynamic Analysis of Hydrogen Carrier Materials

Abstract:

Hydrogen storage by reversible chemical reactions is an attractive option to increase energy density and realize storage at ambient conditions. A group of very interesting storage materials are the so called Liquid Organic Hydrogen Carriers (LOHC). Examples currently under research are toluene, N ethylcarbazole or dibenzyl toluene.

One of the biggest challenges in this approach is the release of hydrogen due to unfavorable thermodynamics of the dehydrogenation reaction. Selection of carrier materials that exhibit small enthalpic difference between hydrogen rich and lean form supports hydrogen release. On the other hand, if the energy level of the hydrogen rich form is not sufficiently below that of the hydrogen lean form, storage becomes unstable and elevated pressures or low temperatures are required during storage time.

In addition to their use as hydrogen carrier for energy storage and transport purposes, LOHCs can be used as transport vectors for hydrogen in the chemical industry. In such a case the carrier material should be selected in a way that the enthalpy of hydrogenation of the transfer agent is less than that of the accepting substance. However, to achieve a good energy efficiencyaich the difference should not be too high. Hence, different carrier materials can be optimal for different processes.

In this contribution the thermodynamic aspects of the selection of LOHC materials are highlighted regarding energy systems as well as chemical processes.

422743 Hydrogen Adsorption in Silicon-Carbide Nanotubes Doped with Potassium and Titanium

Abstract:

Hydrogen Adsorption in Silicon-Carbide Nanotubes Doped with Potassium and Titanium

Seyed Hamed Barghi, Theodore T. Tsotsis, and Muhammad Sahimi[1]

Mork Family Department of Chemical Engineering & Materials Science, University of Southern California, Los Angeles, California 90089-1211

Abstract

Silicon-carbide nanotubes were synthesized by the reaction between SiO vapor as the Si source and carbon nanotubes as the carbon precursor. The resulting nanotubes were then purified with a hot, concentrated NaOH solution in order to remove the amorphous silica from their surface. The purified silicon-carbide nanotubes were doped with either potassium (K) or titanium (Ti). The hydrogen adsorption isotherms of the nanotubes were measured gravimetrically with the aid of a magnetic suspension microbalance. The results indicate that the K-doped silicon-carbide nanotubes exhibit promise as a hydrogen adsorption material, storing three times more hydrogen than the precursor carbon nanotubes. The gravimetric hydrogen uptake measurements of K-doped SiCNTs reveal a significant increase in their hydrogen adsorption capacity, when compared with pure (un-doped) SiCNTs. An increase in the binding energy caused by the charge transfer from the metal impurity to the Si and C atoms is thought to be the reason for the superior performance of the doped nanotubes (The pure SiCNTs, themselves, are also capable of adsorbing a much higher amount of hydrogen, when compared with the CNTs, mostly due to the charge transfer from the Si atoms to the C atoms in the SiCNTs' structure). However, the presence of adsorption sites with increased binding energy in the K-doped SiCNTs gives rise to desorption irreversibility. This means a higher temperature would be required to release the adsorbed hydrogen molecules, which is unfavorable from an energy storage point of view.

Interestingly enough, the Ti-doped SiCNTs show inferior performance than the pure SiCNTs (though still superior to that of CNTs). This contrasts prior molecular simulation studies that indicate that the doping with Ti should significantly improve the sorption capacity of the SICNTs. The reasons for this surprising finding are not entirely clear at this point and time (and should be further investigated), but the behavior is suspected to be due to the presence of an oxygen impurity introduced during the preparation of the materials. The additive hydrogen storage capacity (AHSC) of the pure and K-doped SiCNTs are positive throughout the region of pressures investigated. This is definitely an important improvement over the CNTs, for which their AHSC turns negative for pressures higher than 20 bar, implying that a hydrogen storage system working without any solid adsorbent is capable of storing more hydrogen than a CNT-based system (at room temperature).

Keywords: hydrogen storage, SiC nanotubes, doping, titanium, potassium

---

413713 An Energy System Based on Liquid Organic Hydrogen Carrier

Abstract:

In Germany, the increasing share of unsteady renewable energy generation in the total energy supply makes storage options necessary. One option is to use hydrogen, however not in its pure form but rather stored by chemical reaction in a carrier. The transport and distribution of energy is thus the transport of a liquid, a logistical network that already exists for heating oil or car fuel. There is consequently no need to build a completely new infrastructure.

The lecture describes the present technical readiness level of 6 of this technology and its applications and challenges in stationary or mobile energy storage. This state is contrasted to battery storage.

AUTHOR INDEX

AUTHOR INDEX